# De Gruyter Series on Smart Computing Applications

---

Edited by
Prasenjit Chatterjee, Dilbagh Panchal, Dragan Pamucar,
Sarfaraz Hashemkhani Zolfani

## Volume 2

# Knowledge Management and Web 3.0

Next Generation Business Models

Edited by
Sandeep Kautish, Deepmala Singh,
Zdzislaw Polkowski, Alka Mayura, Mary Jeyanthi

DE GRUYTER

**Editors**

Dr. Sandeep Kautish
LBEF Campus
Opp. Maitidevi Temple
Kathmandu
Nepal
dr.skautish@gmail.com

Asst. Prof. Deepamala Singh
LBEF Campus
Opp. Maitidevi Temple
Kathmandu
Nepal
2udeepmala@gmail.com

Dr. Zdzislaw Polkowski
Wroklaw University
of Economics and Business
Komandorska 118/120
53-345 Wrocław
Poland
zdzislaw.polkowski@ue.wroc.pl

Dr. Alka Mayura
AMITY University
Sector 125 Amity Rd
Noida 201301
India
alkamya@gmail.com

Dr. Mary Jeyanthi
Faculty of Business Intelligence
and Artificial Intelligence
Intitute of Management and Technology
SH 248 35 Km Milestone, Katol Road
Maharashtra 441502
India
dr.maryprem@gmail.com

ISBN 978-3-11-072264-2
e-ISBN (PDF) 978-3-11-072278-9
e-ISBN (EPUB) 978-3-11-072293-2
ISSN 2700-6239

**Library of Congress Control Number: 2021941296**

**Bibliographic information published by the Deutsche Nationalbibliothek**
The Deutsche Nationalbibliothek lists this publication in the Deutsche Nationalbibliografie; detailed bibliographic data are available on the Internet at http://dnb.dnb.de.


Cover image: monsitj/iStock/Getty Images Plus
Typesetting: Integra Software Services Pvt. Ltd.
Printing and binding: CPI books GmbH, Leck

www.degruyter.com

# Contents

# List of Authors

**Dr. Deepmala Singh**
Assistant Professor, Asia Pacific University
(Kathmandu, Campus), Malaysia
Email: 2udeepmala@gmail.com

**Dr. Anurag Singh**
Assistant Professor, School of Business
University of Petroleum and Energy Studies
(UPES), Dehradun, India
Email: anuragonmail110@gmail.com

**Sagar Karki**
Student (MBA), Asia Pacific University
(Kathmandu, Campus), Malaysia
Email: sagar.karki@study.lbef.edu.np

**Afreen Niyaz**
Research Scholar, Institute of Home Science
University of Kashmir, Hazratbal
Srinagar, India
Email: afreeeniaz@gmail.com

**Dr. Vineet Chouhan**
Assistant Professor, School of Management
Sir Padampat Singhania University, Udaipur
Rajasthan- India
Email: vineet.chouhan@spsu.ac.in

**Dr. Shubham Goswami**
Assistant Professor, School of Management
Sir Padampat Singhania University, Udaipur
Rajasthan- India
Email: shubham.goswami@spsu.ac.in

**Dr. Pranav Saraswat**
Assistant Professor,
Nirma University, Ahmedabad, Gujarat, India
Email: pranav.saraswat@nirmauni.ac.in

**Navita Mahajan**
Amity University, Noida, Uttar Pradesh, India
Email: navitamahajan07@gmail.com

**Seema Garg**
Amity University, Noida, Uttar Pradesh, India
Email: seemagarg1@gmail.com

**Oridoye Tomide Hakeem**
Chi Limited, Lagos, Nigeria
Email: oridoymero@yahoo.com

**Pradeep Bedi**
Lingayas Vidyapeeth, Faridabad, India

**S. B. Goyal**
City University, Petaling Jaya, Malaysia
Email: drsbgoyal@gmail.com

**Jugnesh Kumar**
St. Andrews Institute of Technology and
Management, Gurgaon, India

**Ritika**
St. Andrews Institute of Technology and
Management, Gurgaon, India

**Sandhya Java**
Department of Humanities Arts and Applied
Sciences, Amity University, Dubai, U.A.E
Email: sandhyaj@amitydubai.ae

**Hiba Mohammed**
Department of Humanities Arts and Applied
Sciences, Amity University, Dubai, U.A.E
Email: hibam@amitydubai.ae

**Aradhana Balodi Bhardwaj**
Department of Humanities Arts and Applied
Sciences, Amity University, Dubai, U.A.E
Email: abhardwaj@amityuniversity.ae

**Vinod Kumar Shukla**
Department of Engineering
and Architecture, Amity University
Dubai, U.A.E.
Email: vshukla@amityuniversity.ae

https://doi.org/10.1515/9783110722789-203

**Tapash Rudra**
City University, Malaysia
Email: tapashrudra6@gmail.com

**Dr. Purvi Pujari**
Institute of Management Studies and Research, Bharati Vidyapeeth
Email: drpurvipujari@bvimsr.com

**Prof. Priyeta Priyadarshini**
Institute of Management Studies and Research, Bharati Vidyapeeth
Email: priyeta.priyadarshini@bvimsr.com

**Ch. Rupa**
Department of Computer Science and Engineering, Lincoln University College Malaysia
Email: rupamtech@gmail.com

**Divya Midhun Chakkarvarthy**
Department of Computer Science and Multimedia, Lincoln University College Malaysia
Email: divya@lincoln.edu.my

**Deepa Sahai Garg**
Assistant Professor, College of Vocational Studies, University of Delhi, Delhi, India

**Tarakeshwar Gupta**
Visiting Faculty and Research Scholar, School of Management, GD Goenka University Sohna, Haryana, India

**Dr. Anuradha R. Tiwari**
Associate Professor, Economics Department School of Humanities and Social Sciences GD Goenka University, Sohna, Haryana, India

Deepmala Singh, Anurag Singh, Sagar Karki

# Knowledge Management and Web 3.0: Introduction to Future and Challenges

**Abstract:** Today, it is understood that more data is streaming into organizations from more sources than ever before. But only collecting and storing the data is not enough for every organization, it is important to put them in usable form so that they may get easily accessible whenever they are needed. This overburden of information produced new technologies in many forms. In these new technologies knowledge management and Web 3.0 are also used as a tool for the better use of the data in the productivity of the organization. This chapter will discuss that how integration of new emerging technology, that is, Web 3.0 in knowledge management can impact the working of an organization along with that the chapter also explains the evolution and overview of knowledge management and Web 3.0, the future, challenges, and the significance of the same.

**Keywords:** Knowledge Management, Knowledge Sharing, Knowledge Stockpiling, Knowledge Transfer, Web 1.0, Web 2.0, Web 3.0

## 1 Introduction to knowledge management

With the speedy growth of Internet, knowledge has been increasing rampantly year by year. By 2004 approximately every 18 months the knowledge was doubling (American Society of Training and Documentation (ASTD)), with the advent of Web.

The convenience and approachability of Internet and the popularity of computer have produced a major change in the passion and way of using the information, the popularization of using computer has given a massive transformation from traditional web pages Web 1.0 to a new emerging internet model, which is based on ample content and collaboration known as Web 3.0. Organizing the gathered knowledge on the internet by knowing the language of machines and humans without the help of search engines and the recovery of information are the main features of Web 2.0. Today we are in the era of information society which is not only information based but there are ample opportunities of producing and diffusing new knowledge. Since to achieve success for every enterprise knowledge is a core factor in today's highly competitive commercial environmental and that is why designing the knowledge management (KM) adjusted with the evolution trend of internet is one of the important topics for all type of enterprises.

Today every enterprise either big or small has been overburdened with information's by which it has become difficult to separate the useful information from the

https://doi.org/10.1515/9783110722789-001

bulk of given information. The important is to share better and relevant information rather than collecting and sharing more information. Many literatures show that KM 3.0 with the help of Web 3.0 will be a valuable tool in using knowledge rather than gathering and collecting the information [16].

Many researches have been conducted regarding the use of KM and Web 3.0 in various fields, this article aims to discuss how the combination of KM and Web 3.0 influence and give an effective construction of HRM 2.0. This chapter will discuss that how integration of new emerging technology, that is, Web 3.0 in KM can impact the working of an organization.

KM is an interesting issue in business networks. Even though the term "knowledge management" may recommend a somewhat straightforward definition, there are a lot of suppositions on what precisely it is and how it ought to be utilized. Notwithstanding, because of the consistently expanding speed of business advancement, the assignment of successful and serious executives of associations gets fundamental, and KM, whenever comprehended and actualized appropriately, might be a helpful apparatus for business change just as the way to competitive advantage.

Davenport and Prusak [8] see knowledge as a developing blend of outlined insight, values, relevant data, and master understanding that gives a structure to assessing and consolidating new encounters and data. They found that in associations, the information regularly gets implanted in archives or vaults and authoritative schedules, cycles, practices, and standards. They additionally express that with the end goal for information to have esteem, it should incorporate the human increases of setting, experience, and understanding. Nonaka [22] grows this view by expressing that information is tied in with importance in the sense that it is setting explicit. This suggests that clients of knowledge should comprehend and have experience with the unique circumstance, or encompassing conditions and impacts, in which the information is created and utilized with the end goal for it to have significance to them. This likewise infers that all together for an information vault to be helpful, it likewise should store the setting where the knowledge was produced. That knowledge is setting explicit and contends against the thought that knowledge can be applied all around; notwithstanding, it does not contend against the idea of hierarchical information [6]. This part believes hierarchical information to be a vital segment of what authoritative individuals recall and use, implying that knowledge is noteworthy. KM is the cognizant cycle of characterizing, organizing, holding, and sharing the information and experience of representatives inside an organization [5] characterizes information as taking five particular structures: encapsulated, inserted, embrained, encultured, and encoded. He characterizes typified knowledge as knowledge that is acquired through preparing of the body to play out an undertaking [17] characterized KM as the act of specifically applying information from past encounters of dynamic to current and future dynamic exercises with the express motivation behind improving the association's viability.

With advances in computing, the advent of the internet and more specifically the Web, KM tools have developed. It is no longer just a matter of processing documents,

but rather a set of internal and external data so that they can be reused in the future. Here are some examples of tools:
- **Forums or instant messaging:** simplifying communication between different players;
- **Word, Excel, videos, etc.**: Establishing written or oral procedures;
- **Intranet platforms**: sharing news and general documents;
- **Directories:** location of different expertise;
- **Working groups**: developing a collaborative system and sharing among members of the same organization;
- **E-learning training:** optimizing the learning of newcomers and upgrading employees.

**Several features or keywords can help identify the KM's practices:**
- a managerial and/or managerial approach (will, device, measures, etc.) opposed to the idea of letting things happen at random;
- a goal: to do better (than before or others), to win;
- a collective approach: moving from the individual who manages his own knowledge to the collective mobilization of individual knowledge to move toward collective knowledge;
- an inscription in time, in time (memory, sustainability, capitalization, etc.): moving from activism with its short-term "blows" to a thoughtful construction, assured of a future based on the exploitation of the best (and failures) of the past;
- a vital resource to be mobilized, the knowledge(s), or, in the end, the heads of men, their knowledge, their experiences.

## 1.1 Significance of knowledge management

KM is significant because it supports the effectiveness of an association's dynamic capacity. In ensuring that all representatives approach the general expertise held inside the organization, a more brilliant labor force is assembled who are more ready to make brisk, educated choices that advantage the organization. The significance of KM is developing each year. As the commercial center turns out to be perpetually serious, perhaps the most ideal approach to remain on top of things is to construct your organization in a clever, adaptable way. You need to have the option to spot issues in good ways and react rapidly to new data and developments. If for you KM is akin to the vision of a grizzled archivist surrounded by dusty books, then these few lines are for you.

The real challenge is not limited to the acquisition of knowledge, but lies well in his mastery. Because it is essential to identify and manage existing knowledge, to know its veracity, its internal or external criticality, to put in place the best way to collect, sort, disseminate, value it or find it. The KM is undoubtedly a key player in the company. The added value of the company, beyond its capitalization, goes through three fundamental areas necessarily multicultural to respond to an effective globalization:

- Knowledge of its customers and the market
- Knowledge of past, current or up-to-date IT technologies
- The methodologies that enable them to be implemented in industrialization objectives

The KM allows each employee to know and share his knowledge and know-how regardless of his location and his original training. For example, the establishment of virtual skills centers carried by web tools, allows to create a community of men sharing the technical and methodological knowledge of customers through a set of common collaborative tools.

The in-house organization of these skills centers requires synergies between the cross-cutting functions of the company that places the knowledge manager away from the dusty rooms and close to driving change: it requires moving from a way of working based on the power of knowledge to a collaborative way of sharing knowledge.

The goal of the KM is to offer customers increasingly industrialized services while maintaining closeness both to the point of customer knowledge and transparency of the service offered to the customer. Thus, KM is no longer confined to the company but it is a common service that opens a new relationship with the customer, on a common base of information and tools of company content management (ECM). This mastery of collective knowledge allows, beyond the monitoring of the service, an important support to the processes of continuous improvement on the service and transforms the customer/provider relationship into a partnership, bearing the concept of extended enterprise, where end-user satisfaction is at the center. Moreover, the mastery of knowledge and know-how is the guarantee of the mastery of the added value of the company in a context where the modes of communication and information sharing are changing the landscape of IT companies. If our century is indeed that of knowledge, then our industry can only be valued by a constant and visible mastery of its intangible capital. Although this point is now recognized by all, its valuation remains difficult in the markets.

The KM is a major axis of this valuation. Indeed, the mastery of one's knowledge opens up the fields of knowledge. Knowledge inference and expressing knowledge are the two primary kinds of knowledge covered inside the meaning of KM [3]. Implied information is more instinctive, less simple to bundle and impart to other people. Instances of inferred knowledge are creative reasoning and understanding non-verbal communication. Unequivocal information will be data that is effortlessly arranged and instructed, for example, how to change the toner in a printer and numerical conditions. It will guarantee that the specific information on representatives does not leave with them, or go unutilized by different employees who might profit from that information. It takes into account better situational mindfulness, just as opening entryways for finding out about prescribed procedures, and in general hierarchical improvement.

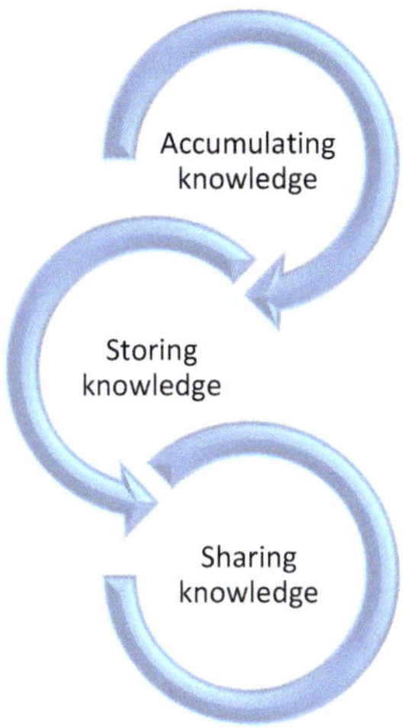


**Figure 1:** Three key areas of knowledge management.
Source: Authors Compilation

## 1.2 Knowledge management: an organizational tool

The need to adjust KM methodology to organization technique was distinguished as basic to the achievement of KM [12, 23]. An organization's KM technique should make a comprehension of the institution's KM assets and where they live; articulate the job of knowledge in worth creation, and contain various coordinated undertakings or exercises staged after some time including brisk successes just as long-haul benefits [9]. An institution's methodology of KM is not discretionary yet depends on the way the organization serves its customers, the financial matters of its business, and the individuals it employs. The question is no longer whether or not, knowledge is a critical organizational resource, but on the alignment of KM to corporate strategy, measurement of KM impact, and driving of business results.

## 1.3 Knowledge stockpiling

Knowledge stockpiling includes gaining knowledge from individuals from the organization or outer source [15]. Associations are likewise failing to remember while making knowledge and learning. Along these lines, the protection, association, and organization of knowledge that is the organization's memory establish a significant and successful piece of authoritative information. Aggregate or authoritative memory is characterized as past information, experience, and occasions. Hierarchical memory incorporates information that exists in various structures, including recorded reports, organized data put away in electronic data sets, information on coded individuals put away in master frameworks chronicled authoritative techniques and cycles, and organizations of people and people that learn implied information [14]. Authoritative memory is partitioned into semantic or plots. Semantic memory alludes to general, unequivocal, and expressive information, while plot memory alludes to information in explicit settings and settings. The change of information into semantic memory does not ensure the compelling spread of information, nor does it ensure proficient capacity.

## 1.4 Knowledge transfer

Knowledge transfer exists at various levels. The exchange of knowledge between people, from individual to group, between groups, across groups, from group to association [4]. Considering the appropriated idea of authoritative agreement, a significant cycle of KM in a hierarchical setting is the exchange of information to where it is required and where it is utilized. In any case, this is not a straightforward cycle since organizations do not have a clue what they know and there is no incredible framework to find and recover information put away in the association. Information transfer comprises five components, to be specific, perceiving the estimation of the information on the source individual, the motivation inclination of the source, the presence and advancement of the correspondence channel, the motivator propensity of getting the individual, and the absorptive limit of the accepting individual [26].

## 1.5 Knowledge implication

The upper hand of a venture is gotten from the use of information instead of the actual information. It is implied that information itself cannot deliver the upper hand. Simply by utilizing information and applying the part of information can the benefit of the undertaking to be created. The information reconciliation system that makes authoritative capacities is the guidance, association, and free team. The marker alludes to the particular guidelines, principles, methods, and direction shaped by the change of unequivocal information and incorporates information that is adequately spoken with non-specialists through master unspoken knowledge. Authoritative undertakings allude to the type of assignment execution and collaboration, intuitive conventions, and preparing details that permit the application and coordination of individual information without the need to discuss their insight with others. Groups with the essential information and aptitude are joined to take care of issues when task vulnerability and multifaceted nature forestall guidance and hierarchical exchange details.

**The challenge of the KM is to be an engine of innovation in two aspects:**

- The financial enhancement of technical innovation with the patent deposit and technical or methodological innovation.
- The mastery of his knowledge and know-how that allows an inter-technological and methodological sharing more important than before.

**One of the foundations of innovation is the use of technologies in other fields than its original domain.** The universal science of the enlightenment went toward a specialization that was growing until the end of the last century. The advent of the Internet and a number of sciences brings to light a new vision that carries the silo vision of knowledge back to other times, to propose a heuristic, "web" vision that allows innovative rapprochements that must be followed and supported. Patent filing is the logical conclusion that can only be made with a constant mastery and monitoring of knowledge within the company.

**KM is a central player in the company,** less greying than it appears. This form of management is at a crossroads between the innovative technophile, the savvy ECM tool user, the ethnologist of the IT market and his own company and the historian of a changing corporate culture at the service of men for the most efficient service to the customer.

# 2 Knowledge management issues

Increased competition and the management of unforeseen events are pushing organizations to develop their responsiveness. To do this, they must necessarily complement automated systems by the intervention of man and his knowledge. If the direct challenge of KM is to control knowledge within an organization, it results from other strategic issues.

# 3 Future of knowledge management

KM creates an environment conducive to innovation through the constant improvement of skills. It is the result of careful structuring of the organization and rigorous application of procedures in order to increase the quality and speed of the work.

The time savings, the increase in the skills of the employees, and the transfer of knowledge have many positive impacts:

- Accelerate the learning capacity of newcomers
- Promote decision-making by reducing subjectivity
- Increase team productivity
- Increase customer satisfaction

- Adapt more easily to changes (changes in the market, departure of an employee, etc.)
- Engage and motivate the work teams

Thus, the sharing of expertise at work directly affects the productivity of the company. A productive, responsive, and innovative organization will then be more competitive and more valuable.

# 4 Introduction to Web 1.0

Technology has improved a lot in many areas of human life and the same improvement has occurred in the internet industry. The news of the launch of the Web 1.0 is not so old since it was first launched in the year 1990, and in 1991, the first-ever website was launched in the world of the internet. Web 1.0 is commonly known as the first interphase in the website world. Web 1.0 was also known as the era of a new information world where people can read and get the information directly by visiting the source. Web 1.0 is also known as a read-only web service, where users can visit the website and read the information without making any interaction or comment for the provided content. In this era, the websites were just simply coded and designed to provide information to a user for reading purposes only, and the users were not able to make any comments or provide reactions if any user wished to do it. Moreover, this system is purely designed in the favor of website owners, since they are the ones who can control the information which they want to provide for their users at any time [13].When the World Wide Web was started, the websites which were created for users were very basic and simple. This stage used to know as the first stage of the website where users can only read static websites. People were not aware from the full content information, and their primary requirement was minimum from websites, which was sufficient for the users. The websites during 1.0 were offering common information-sharing space to users which helps users to understand brochures details, helps them to locate phone numbers, email, fax details, and addresses online at any time for everybody. This system was mainly based on the protocols of XML, HTTP, and HTML programs [25].

# 5 Web 2.0

The technology was changing frequently in every field, and the same thing has happened in the world of the internet as well. People have now realized that they need more advanced technology and facility in the website world too. It was the time that users were not only convinced for the read-only facility, they were also thinking to update the information if they find any necessary changes. After understanding this basic requirement, Web 2.0 was introduced in the technological world. This second-generation technology in the web system was officially introduced in October 2004 in the internet market. The best feature about Web 2.0 was, in this system, the website had the additional facility of interaction capabilities and it got widely popular as read and write web. There are many popular examples of Web 2.0 websites and each one of us is still using them in our daily life such as different blog websites, social media, and popular sites like Facebook and YouTube [20]. It helps the user to manage and assemble their information-sharing habits and link with the different web applications which were connected with World Wide Web platforms. The biggest example has got the boom in their platforms since more and more people started to use those sites in their regular life. Web 2.0 was the absolute successor of Web 1.0 since it was offering more user-friendly interactive platforms which were useful for the creation of collaborative content, updates, and create flexible web designs. This also helps to introduce new services like chats, playing games, e-learning, blogging, communication, and freedom of tag and comments on anything which are possible to share with everyone. Web 2.0 was more interaction-based with loaded with more web applications unlike Web 1.0, which was only based on information sharing based on static one-way content web forms [1].

# 6 Web 3.0

Now let us talk Web 3.0, and every individual needs to understand what 3.0 is and what it is capable of. When users search the information on Web 1.0 and Web 2.0, then the system was searching and presenting the information via using the keywords which were searched by the user. However, this outline has changed after the launch of Web 3.0. With the launch of this technology, the machine learning was also included in Web 3.0 which is widely popular as artificial intelligence (AI). Web 3.0 was the semantic web which was combined with read-write-execute the information by the user, where computers and servers can interpret the searched information like humans with AI and machine learning methods. In this system, the algorithm of the programs was specially designed to create communication between machines to the machine. This is the third-generation technology in the field of the World Wide Web which was launched in 2006 by creating this technology with semantic web data integration. Web 3.0 is using the technology with metadata, where it converts the display only data to meaningful information which helps the users to evaluate their searches. The semantic Web 3.0 is the read-write intelligent web feature that is developed by the smart applications to provide users with highly portable and individual-oriented data with next-generation browsing and information searching capabilities while searching for content. This system was executable semantic-based which used the AI and multi-user virtual environment to provide users a better experience when it compared with Web 2.0 which was only based on reading and writing program with online social networks platforms. This feature enables the internet program to help this system to rethink themselves and generate the results rather than humans [7].

After the launching of Web 3.0, it got the hype quickly due to its application services which helps search engines to produce and show more complete, reliable, and targeted information that users want to see in their search. Not even that, it has also helped the people to tightly connect with their personal and professional life and helps them to manage their businesses and friend circle simultaneously. Apart from this, Web 3.0 also helps users to provide a secured maintained record and secured place to store their information digitally so that the user can access it whenever they want. Unlike Web 2.0 where users were only able to connect with the social networks and to share their views, now users can link their feelings, thoughts, and ideas too which is the evolving extension with AI for users in regarding search terms on search engines [2].

Well to be more precise, every technology has its advantages and disadvantages. During my research, I have figured out a couple of advantages and disadvantages of Web 3.0, which will help any user to understand the pros and cons of this technology. Let us start with the advantages first and it starts with increased information linking. This program is based on the semantic-based program where the search engine search feature helps connect data points across the web. Web 3.0 is more efficient when it comes to the matter of web browsing by making communication a breeze while using social network websites. The interaction among people also increased and was spearheaded by faster internet connections, unlike the previous Web 2.0 or Web 1.0 versions. This Web 3.0 makes online interactions easier than offline. Let us move out to the major drawbacks of Web 3.0 and the first one I would like to state is less advanced devices. Web 3.0 has improved a lot of its functions but unless and until if different individuals do not have the devices, they cannot be able to handle Web 3.0 features use those facilities. The complexity increases among users where it can be complicated for newcomers to understand and adapt to the Web 3.0 implementation challenges. The vastness in Web 3.0 needs to eliminate all semantically duplicated terms from billions of pages on the huge internet [21]. The user quarries are not specific and it sometimes can be extremely vague to understand. Web 3.0 also has to deal with scores of uncertain value and deceit the provided data which can lead to logic. To improve all these issues, the semantic blockchain technology could help to interconnecting and storing data in a secure with smart contracts. By using decentralized ledgers in the form of decentralized apps, and using different applications can run on their own without central points of failure in a server farm. It helps track specific information and connects it to other apps responsible for other information. The most interesting part of this development is that nobody knows how long it will take to reach a fully functioning Web 3.0. One thing is sure in Web 3.0, certainly, Web 3.0 will change our online lives if it has improved further and combined with AI to produce and provide more accurate results to its users. The future of Web 3.0 is surely mind-twisting and exciting.

After Web 3.0, it is argued that Web 4.0 is going to launch, however, it will be too early for the individuals to talk about or expect Web 4.0 in the internet market because the idea behind this is still in progress and undergrounded. The biggest thing is there is no significant information or definition discovered yet. However, the major computer science experts believe that the upgraded version of AI is going to be launched soon since many companies are working to enhance their AI and machine learning responses into computer applications. Computer experts also believe that, after upgrading AI in a more precise way, the human and machine can interact and communicate with each other more smoothly to think and make decisions on commands [22].

## 6.1 The four key properties of Web 3.0

1) Semantic web – it is the key for Web 3.0 making the data easy to process for machines. It is helping understand the meaning of words to generate, share and connect content through search and analysis. It evolving the user experience to another level of connectivity which leveraging all the available information available online.
2) AI – AI is one of the major key factors regarding the popularity of Web 3.0. It enables machines to become more intelligent with the vast web data to satisfy the needs of the users. AI allows websites to filter and present the user with the best data possible.
3) 3D graphics – Web 3.0 has surpassed its predecessors due to its three-dimensional technology where it provides a more realistic three-dimensional online world than 2D. 3D is the base for users' virtual identities which provides a much better experience from Web 1.0 and Web 2.0
4) Ubiquitous – the idea of existing or being everywhere at the same time. The rise of mobile devices is increasing day by day and it is easier for many individuals to access the internet anytime anywhere. This internet access makes Web 3.0 more omnipotent and user friendly.

# 7 Future of Web 3.0

Web 3.0 is blockchain oriented and that means no central point of control, complete data ownership, less interruption of services in permission-less networks. There are myriad reasons regarding the popularity of Web 3.0 in the market and the main reason is their build on a better application system where it will create and produce better-targeted results [10].

# 8 Conclusion

We now know that knowledge and knowledge represent a fundamental intangible capital for the company. With the advent of the web and it, it is possible to update them regularly. KM is then part of a continuous improvement process. A company's performance is no longer valued solely in terms of production. It is more its ability to react in the event of a problem, change or competitive development that shows its performance. Workers are no longer seen as mere enforcers. They hold the knowledge and know-how that constitutes the true wealth of an organization. Beside KM Web 3.0 is a term much more used to talk about the future of the Internet than the concrete technical innovations of recent years. At a time when many research sectors offer innovative applications related to the web and the Internet, trying to group them all under one banner seems impossible. If Web 2.0 has finally managed to bring together ideas and identified uses of all, this is not the case with Web 3.0. The technology race of the past 15 years has led to extraordinary innovations, both online and within our software and hardware. A problem then arises: these uses are all linked to the Internet, but not always with the Web.

# References


[1] Dwivedi, A. & Kumar, S. (2011). Current security considerations for issues and challenges of trustworthy semantic web. International Journal Advanced Networking and Applications, 3(1), 978–983.

[2] Aghaei, S. & Nematbakhsh, M. (2012). Evolution of the world wide web: From Web 1.0 to Web 4.0. International Journal of Web & Semantic Technology, 3(1), 1–10.

[3] Bezzina, F., Baldacchino, D. & Cassar, V. (2020). Relating knowledge management enablers, knowledge management processes, and organizational effectiveness. International Journal of Knowledge Management, 16(4), 109–124. doi: https://doi.org/10.4018/IJKM.2020100106.

[4] Bloodgood, J.M. & Salisbury, W.D. (2001). Understanding the influence of organizational change strategies on information technology and knowledge management strategies. Decision Support Systems, 31(1), 55–69. doi: https://doi.org/10.1016/S0167-9236(00)00119-6.

[5] Blackler, F. (1995). Knowledge, knowledge work, and organizations: An overview and interpretation. Organisation Studies, 16(6), 1021–1046.

[6] Courtney, J.F. (2001). Decision making and knowledge management in inquiring organizations: Toward a new decision-making paradigm for DSS. Decision support systems, 31(1), 17–38.

[7] Cohen-Almagor, R. (2011). Internet history. International Journal of Technoethics, 2(2), 45–64.

[8] Davenport, T.H. & Prusak, L. (1998). Working Knowledge, Boston, MA, Harvard Business School Press.

[9] Du Plessis, M. (2007). Knowledge management and what makes complex implementation.

[10] Dwivedi, A., Kumar, S., Dwivedi, A. & Sing, D.M. (2011). Current security considerations for issues and challenges of trustworthy semantic web. International Journal Advanced Networking and Applications, 3(1), 978–983.

[11] El, L.E. (2009). A brief history of the internet. Computer Communication Review, 39(5), 22–31.

[12] Gao, F., Li, M. & Clarke, S. (2008). Knowledge, management, and knowledge management in business operations. Journal of Knowledge Management, 12(2), 3–17. doi: http://10.1108/13673270810859479.
[13] Hiremath, B.K. & Kenchakkanavar, A.Y. (2016). An alteration of the web 1.0, web 2.0 and web 3.0: A comparative study. Imperial Journal of Interdisciplinary Research (IJIR), 2(4), 705–710.
[14] Ha, T.M., Kühling, I. & Trautz, D. (2020). A systems approach toward climate resilient livelihoods: A case study in Thai Nguyen province, Vietnam. Heliyon, 6, 11. doi: https://doi.org/10.1016/j.heliyon.2020.e05541.
[15] Holsapple, C.W. (2001). Knowledge management support of decision making. Decision Support Systems, 31(1), 1–3. doi: https://doi.org/10.1016/S0167-9236(00)00115-9.
[16] Interpretation. Organisation Studies 16(6),1021–1046.
[17] Iversen, A. (2009), KM 3.0 Part IV: A practical KM system, http://www.ppcsoft.com/blog/km-3-4.asp
[18] Jennex, M.E. (2005). What is KM? International Journal of Knowledge Management, 1(4), i–iv.
[19] Knowledge Management, 16(6), 862–878, doi: 10.1108/13673271211276164
[20] Singh, K.P. & Gulati, D. (2011). Technological march from web 1.0 to web 3.0: A comparative study. Library Herald, 49(2), 146–157.
[21] Lal, R. & Lal, M. (2011). Web 3.0 in education & research. BIJIT – BVICAM's International Journal of Information Technology, 3(2).
[22] Nath, K. & Iswary, R. (2015). What comes after web 3.0? web 4.0 and the future. International Conference on Computing and Communication Systems.
[23] Nonaka, I. (1994). A dynamic theory of organizational knowledge creation. Organization Science, 5(1), 14–37.
[24] Oluikpe, P. (2012). Developing a corporate knowledge management strategy. Journal of Knowledge Management, 16(6), 862–878. doi: http://10.1108/13673271211276164.
[25] Patel, K. (2013). Incremental Journey for World Wide Web: Introduced with Web 1.0 to Recent Web 5.0-A Survey Paper. International Journal of Advanced Research in Computer Science and Software Engineering, 3(10), 410–417.
[26] Rubenstien- Montano, B., Liebowitz, J., Buchwalter, J., McCaw, D., Newman, B. & Rebeck, K. (2001). A systems thinking framework for knowledge management. Decision Support System, 31(1), 5–16. doi: http://www.ppcsoft.com/blog/km-3-4.asp.

Afreen Niyaz

# Impact and Role of Knowledge Management and Web 3.0 in Education Sector: An Empirical Study of University of Kashmir

**Abstract:** The process by which knowledge is created, shared, and used is termed as knowledge management (KM) and it is all about changing the way everybody works. The KM is promoted as a significant factor for the organizational sustenance as well as for the upkeep of the worthwhile strength. Intensive discussions have been carried out regarding the importance of KM within our society since past so many years. As we are aware that different technologies have evolved over a period of time, thus reducing the time required for disseminating the knowledge and they are even allowing student autonomy, greater collaboration as well as improved academic efficiency. These applications have mushroomed and at the same time social media has even attracted considerable attention from students of different fields. The popularity of such technologies is influencing the means by which different libraries and other cultural heritage organization's function. The continuous evolution of the innovative internet has unlocked unimaginable opportunities in web-based education. The present study aims to investigate the usage of Web 2.0 tools and the level of awareness among the Post Graduate (PG) students about Web 3.0 technologies at the University of Kashmir, situated in District Srinagar of Kashmir division. The present study has implemented a descriptive survey method. A self-designed questionnaire was used to gather the required information from 200 respondents. The results obtained showed that the PG students of the Kashmir University are accustomed with most of the tools of Web 2.0 like Instant Messaging, SNS, wikis, and blogs while, they are not acquainted with Podcasts, social bookmarks, or RSS feeds. The study also revealed that the most commonly used tools include WhatsApp, Facebook, YouTube as well as Wikis. The respondents specified interacting with friends and family, sharing ideas for their individual life and connections with new people as the topmost reasons for linking with social networking sites. It was found that the PG students are not fully aware about the latest Web 3.0 technologies and need more guidance in that very field. It is suggested that academic institutions should integrate units related to Web 2.0 as well as Web 3.0 tools into the courses in order to train the professionals of future by improving their proficiencies in order to be competent for the future challenges of using these tools.

**Keywords:** Internet, Learners, Technology, Knowledge Management, Kashmir University, Social Network

https://doi.org/10.1515/9783110722789-002

# 1 Introduction

In recent times, we have observed the availability of surplus online services among numerous universities and educational institutions which includes procedure for online admissions, virtual (online) learning environments, and online fee payments in order to facilitate and enhance the lifelong learning among variety of students and to make this compatible with other educational management activities as well [5]. Moreover, a teacher now usually creates a purely web-based delivery system which includes online handouts in respect of student's performance and activities as well as the list of different resources for the reference. We are aware that Web 3.0 is more than just a set of effective and useful new technologies or services. It offers a wide collection of services to make a true online classroom a big real thing. These services have a very positive and real impact on the overall teaching learning process. Such benefits can straight away be aligned to the existing best practices in the online education. Moreover, it is expected that the advantages of this system would be surely aligned to the existing best practices happening in online education and also make further authenticated and resourceful educational environment. The unceasing evolution of information and communication technologies (ICTs) is a doorway to new and innovative opportunities, reforms or improvements in the realm of technology-based education [3]. The current version known as Web 3.0 consists of a more interactive as well as dynamic interface with extremely high advanced features that empower machines to not only think but also to execute the tasks, while providing them with analysis and interpretation skills in order to suit the needs of these users [16]. Hence, it is very important and productive to study and understand the evolution of education technology along with the enhancements that ought to be with Web 3.0 including the way it facilitates teaching and learning process in an advanced manner as well as the limitations that have to be considered and the problems to be addressed.

# 2 Literature review

The technology has got robust advances, since the development of World Wide Web and all the network of documents has advanced into variety of databases. Such a progress is an outcome of three stages of internet revolution [7]. The phase 1 is called the Web 1.0 which involved a single way communication among the user and the website. Another stage is called as Web 2.0 which started during beginning of 2003 [6]. It offers ability to write or we can say two-way communication is involved [39]. The stage 3 of web, that is, Web 3.0 is a notion of an intellectual web which allows user to not only read but at the same time write and execute web as well [40]. This generation involves a cohesive web experience. The internet indeed has performed a dynamic part in the advancement [14]. It is permitting immense student autonomy, improved pedagogic efficiency, and greater collaboration [26]. These applications have grown so much and thus, social media platform has involved extensive attention from students from diverse streams [13]. Such technologies have now dominated the day-to-day professional as well as personal life of so many users across the world because they are so widespread [12]. Another study revealed that the utmost widespread tool-application areas linked with the web comprise, blogs, vodcasts, social sharing services, folksonomies etc. [19] It has been found that these technologies are so important as libraries can plan very attractive facilities using such applications without there being a need to spend extra budgets for online storage [38].

The third generation simplifies the world-wide data warehouse where from various formats of information are gathered and even shared as well [4]. Web 3.0 also includes three dimensional (3D) components which are its key features [37]. Web 3.0 is referred to as the application where data in a website can be interrelated with the associated information of other websites [10]. Another study suggests that the Web 3.0 is where it is the computer that produces new information instead of humans [46]. It has however been recognized that the machineries can perform tasks with users in the content creation process. Various studies explain the features, applications, and technologies that are linked with Web 3.0. [35]. Moreover, the referenced engines would focus on the specific habits along with the preferences of users [9]. In doing so, it will generate more targeted and complex information which would be of great use [34]. Later this information used on a recommendation engine will be gathered and then stored in a manner which is hierarchical in nature by Intelligent agents [41]. Meanwhile such achievements would give Web 3.0 the capacity to not only gather and analyze but at the same time further distribute the data that can ultimately be transformed into tremendous knowledge, useful information, and immense wisdom [18].

The key elements include:

- The ability of new programming languages to categorize and manipulate data in order to allow the machines in order to interpret data and also the phrases describing data [2].
- Information gathering ability from a much wider variety of sources [48].
- The capability of storing and obtaining contextual information from a web search in a hierarchical manner [20].
- The ability to generate as well as share all types of data by all types of devices [28].

The uses of both Web 3.0 and Web 2.0 tools are increasingly being adopted at different higher learning institutions globally in order to support the learning practices as well as the facilitation [17]. The launch of this useful semantic web has indeed resulted in a huge turning point about how the digital native might take lead to take advantage from different kinds of personal learning environments, educational software, or even the personal virtual environments [43]. These kinds of environments permit the digital natives the probability of obtaining meaningful data, data filtering and collaboration to suit their specific needs and requirements [36]. A huge number of higher educational institutions are now beginning to understand and interpret the various new techniques and methods of teaching as well as learning, thus leading to the change in educational institution. All this is really needed to fulfil the ever-changing needs of all the citizens of these information-age societies [42].

# 3 Objectives

- To identify the level of awareness regarding Web 2.0 tools among PG students of Kashmir University.
- To recognize the degree of usage of Web 2.0 tools by the PG students of Kashmir University.
- To find out the level of awareness among the PG students regarding Web 3.0 technologies.
- To examine the reasons for joining the social media platforms by the PG students.
- To discuss the role of knowledge management (KM) in the education sector.

# 4 Methodology

The present study aimed at inspecting the awareness level and usage of Web 2.0 and Web 3.0 tools among PG students of various departments at the University of Kashmir, Srinagar, J&K. The sampling technique used for the study was simple random sampling and the sample size comprised of 200 respondents in number. A self-made questionnaire was designed to gather data from the students studying in different departments of the University of Kashmir. Further, the descriptive statistics were then analyzed with the use of frequency counts as well as percentages. The results that were found were then displayed in the form of tables and diagrams.

# 5 Results and discussion

## 5.1 Awareness regarding Web 2.0 tools among students

The present study which involved 200 participants included half of the respondents, that is, (50%) students who were females and 50% being male respondents. The respondents were enquired to show the extent to which they were aware of the tools of Web 2.0 (see Figure 1). Although, students specified that they were not quite acquainted with Web 2.0 tools as their percent was very low which included RSS feeds, podcasts, and social bookmark. The results showed that the PG students of Kashmir University are accustomed with some tools such as Instant messaging, Social networking sites, wikis, and blogs but they are not enough conversant with tools like RSS feeds, social bookmarks, and podcasts. Therefore, the PG students need to be encouraged for using additional applications to gain enough knowledge and remain updated.

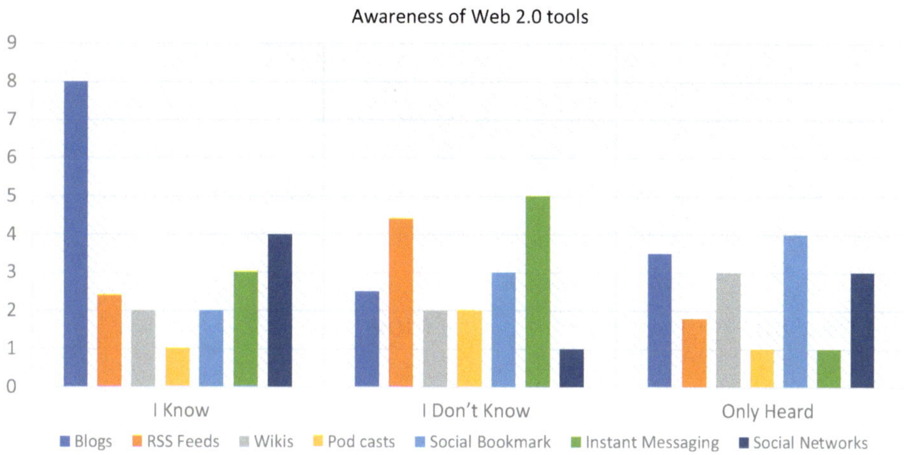


n= 200

**Figure 1:** Awareness of Web 2.0 tools among PG students.

## 5.2 The extent of usage

Respondents were also requested to specify the range to which they were using tools of Web 2.0 (see Figure 2) where it was found that the majority of the respondents (57%) spent maximum time using the social media apps including Facebook, while as 37% respondents were found using Instant Messaging more frequently, 12% respondents were more preoccupied with YouTube platform and used it more often than the other apps for watching videos. In terms of using wikis, 20% of the total participants used it excessively while as 14% spent more time listening to podcasts. RSS feeds and Tagging tools of Web 2.0 were used by a smaller number of respondents and for very less time as they were not quite familiar with these.

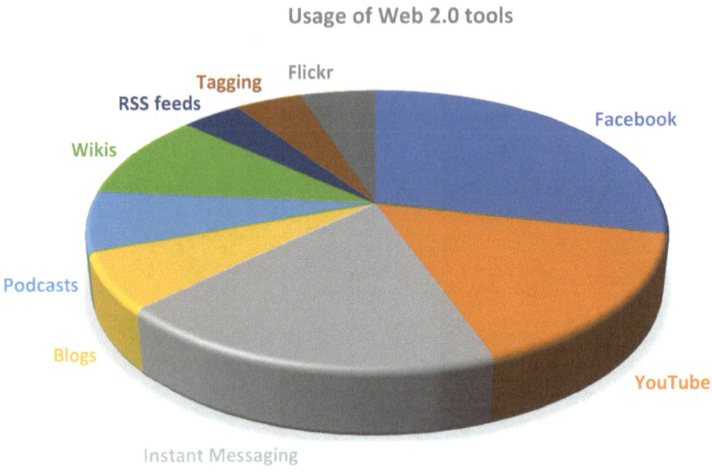


**Figure 2:** Extent of usage of Web 2.0 tools by students.

## 5.3 Reasons for making use of the tools

The next question explored the reasons why PG students of the university joined social networks platforms (see Figure 3). To connect with their friends and family, exchanging views for their personal life or meet new people were found to be the top most reasons for linking with the social networks among majority of the respondents. Also, most of them used these to keep themselves updated. It was also found that only a smaller percentage of students were making use of the SNS for sharing knowledge regarding studies and having interest to know new things with 25.5% and 30% respectively. The findings of the study suggested that the PG students do not use the Web 2.0 tools for only restricted to theoretical drives.

## 5.4 Awareness regarding Web 3.0 among PG students

Respondents were asked if they were aware about the latest web platform termed as Web 3.0 and the response is represented in the Figure 4. Majority of the respondents (65%) were unaware about Web 3.0 applications and did not have any idea of what exactly Web 3.0 refers to and while as 35% of the respondents were fully aware and had enough knowledge regarding Web 3.0 tools and its applications. The PG students having knowledge about Web 3.0 were specifically from the library science background and probably that being the reason they were enough knowledgeable about this new platform of Web 3.0.

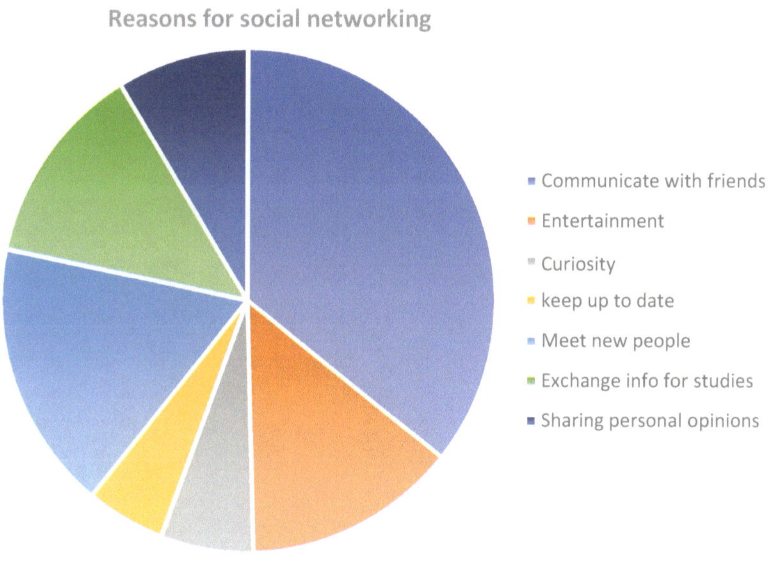


**Figure 3:** Reasons for joining social networks.

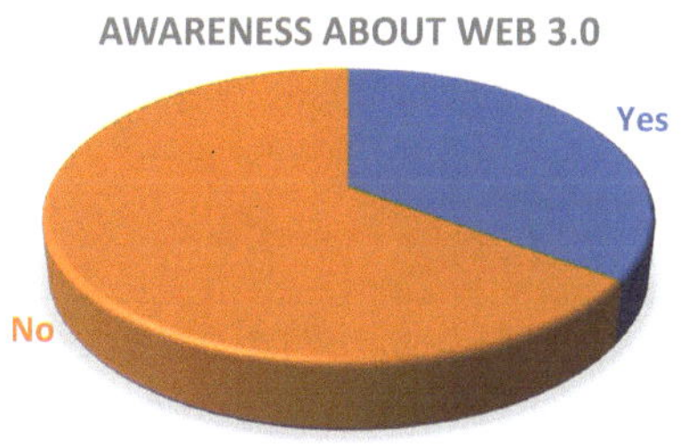


**Figure 4:** Awareness regarding Web 3.0 among PG students.

## 5.5 Knowledge of Web 3.0 tools among PG students

Respondents were asked to indicate the level of awareness regarding Web 3.0 tools (see Figure 5). It was found that most of the respondents (28%) had knowledge about the virtual labs followed by 22% respondents who were mindful of online 3D games. Regarding semantic blogs, participants comprising less percentage, that is, 10% were aware about it and 12% had knowledge of virtual 3D encyclopedia. Meagre percentage of participants was found being aware about the rest of the Web 3.0 tools including intelligent search engines, 3D wikis, and semantic forums. There seems to be a huge knowledge gap among PG students regarding these latest Web 3.0 tools and applications. Educational institutions and organizations need to incorporate the teaching and learnings based on these applications so that students can really benefit from them and stay updated.

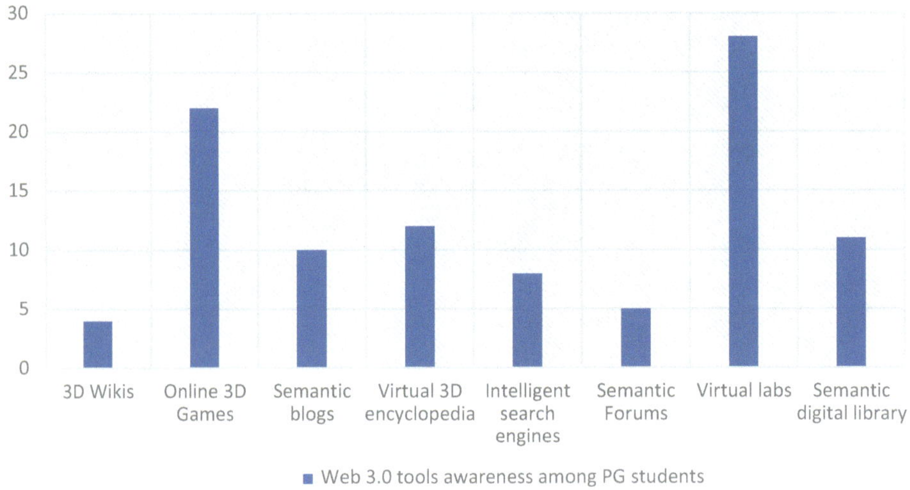


**Figure 5:** Awareness of Web 3.0 tools among students.

# 6 Role of KM in the education sector

An idea that has emerged explosively in different facets of life in the past decades is the KM and the evolution of this KM in the present information era leaped from the understanding that organizational knowledge is a big asset that truly needs to be sustained, nurtured, shared as well as disseminated all across an organization or institutions in order to obtain and also provide solutions to the different organizational problems and concerns [24]. The various techniques and technologies which are specifically used to establish, organize, and build on business expertise and knowledge is known as KM [47]. The concept of KM is related to establishing an explicit knowledge through an organizational and organized process for attaining and governing the flow of knowledge of the employees [32] and this is indeed to make sure that organizations or institutions must achieve their targeted mission as well as vision statement. KM has three important components which include: people, process, and technology. The main instruments of every society are the educational organization for the constant pursuit of knowledge and the essential part of KM in the educational institutions is very important. In other words, we can say the main function of educational organization is the KM [11]. Moreover, through the application of organized thinking, KM in educational organization can surely be uniquely identified as well as studied at various administrative, teaching and learning, research process, student service and lastly the human resource subsystems [21]. The use of the such systems and principles in various educational institutions can develop more flexibility into decision-making processes, easy access to scientific resource, promotion in teaching and learning processes, establishment of the effective internal and external communication network, synergy at the students and faculty knowledge and improvement in quality as well as amidst the research activities in different educational institutions [25]. An important and intrinsic feature of KM is the creation and the development of knowledge in an effective way [30, 8, 33] and this creation of knowledge is very essential for the survival and maintenance of any organization or institutions especially the educational ones. We are aware that knowledge creation is an activity which occurs throughout daily activities, either at work or at the social settings. However, it occurs in different form that could be through humanistic means like talking with people or other formal trainings or even through the technical mechanisms, for example, data mining [22]. It is essentially a human process perhaps technology can facilitate creation of knowledge but it cannot replace people [23]. Mostly organizations have a greater influence over their ability to innovate, create knowledge, and generate value with the new knowledge and it is this knowledge which leads to innovative ways of personal and professional improvements including strategic decision-making capabilities [27]. Moreover, it is very essential for any organization to have the competency to create knowledge as well as generate a competitive advantage that wishes to remain sustainable [15].

The reasons for use of KM are as follows:
- For the productivity of organization
- To promote teamwork as well as collaboration
- To encourage and improve knowledge sharing
- For effective as well as timely services in management
- To develop and boost innovation
- To guard the important knowledge from being vanish
- To encourage outstanding positive work environment
- To endorse critical thinking among employees
- To boost competitiveness in the organization

Indeed, the organizations incorporating KM would surely flourish and achieve its targeted goals in a very effective and efficient manner. It would prove beneficial for not only the organization but for the community as a whole [44].

## 6.1 Functions and influence of KM activities in educational systems

| | |
|---|---|
| **Research activities (technical subsystem)** | – Facilitate interdisciplinary research.<br>– Increase accountability and competitiveness of researches.<br>– Reduce research time and cost.<br>– Improve the quantity and quality of the studies through linking researchers into electronic resources, databases, databanks, etc.<br>– Facilitate the implementation of research based on cross-cultural dynamism in universities as well as to help in the collection of data and information gathering etc. |
| **Educational planning and curriculum development** | – Improve and update educational planning rapidly.<br>– Improve students' accountability and faculty by making use of past experiences.<br>– Encourage quality of different educational programs by applying best practices and also monitoring outputs.<br>– Expand administrative services related to educational processes. |
| **Administrative services (administrative subsystem)** | – Support management processes.<br>– Improve facilities for the students and faculties.<br>– Improve services to internal as well as external stakeholders.<br>– Enhance capabilities in identifying activities.<br>– Attempt to remove the centralization in providing services and connect all sections into knowledge resources.<br>– Increase efficiency and effectiveness of services provided. |
| **HRM (human resource subsystem)** | – Strengthen communication that is informal in nature so as to achieve set university goals.<br>– Prepare as well as update human resources.<br>– Establish a reasonable system of salary and bonus crosswise the organization.<br>– Facilitate responsiveness capabilities and effective communications. |
| **Students activities** | – Create a context of social growth for students from various cultures and communities by providing cyber climate for their discourses.<br>– Improve cooperative activities among students by providing the socio-technological factors for sharing of experiences and knowledge.<br>– Generate an extremely flexible environment for the overlapped learning. |

(continued)

| | |
|---|---|
| **Structure subsystem** | – Lead to integration of KM concepts in the academic sphere to enhance the efficacy of external alliance partnerships and to increase the output of organizational processes.<br>– Preserve organizational resources by improving the knowledge within the organization and utilizing that knowledge for teaching and learning purpose.<br>– Combine both explicit and tacit knowledge shared by staff and faculty through KM. |

# 7 Conclusion

It is concluded with the remark that this study provides information on PG students' familiarity and their usage of tools related to Web 2.0. The study revealed that the PG students of University of Kashmir are accustomed with Web 2.0 tools such as SNS, blogs, wikis, and Instant Messaging. On the other hand, the results also revealed less awareness about the Web 3.0 tools and applications among the PG students of University of Kashmir. Web-based collaborative learning is the most promising learning model in this era of digital learning and moreover the advent of these latest technologies has provided many innovative ways for efficiently building learning groups. Now these groups can easily enable learners to discuss, share, and give feedback to others through the mode of online medium. Also, it permits learners to develop not only their problem-solving but at the same time their critical thinking skills as well. The present paper also discussed about the role of management of knowledge in higher education. It surely benefits the educational institutes in order to enhance their capacity of information gathering and also apply these to problem solving techniques and even encourage the research and frequent upgrading of their work. KM of the educational system must reflect and encompass information at all levels starting from the level of management to the student level in order to develop professional knowledge of the employees. Since it manages large data thoroughly, therefore it can prove to be an extremely powerful tool to augment productivity and decrease cost in the gathering of an enormous volume of data. Moreover, it is very problematic to record tacit knowledge created by any institutional staff. Therefore, if KM practice is being functioned in all institutions as a constant activity only then the knowledge generated could be apprehended as well as preserved for the future use.

# References


[1] Aharony, N. (2009). The influence of LIS students' personality characteristics on their perceptions towards Web 2.0 use. Journal of Librarianship and Information Science, 41(4), 227–242.

[2] Bawab, H. (2014) "Effects of Web 3.0 in the New Digital World". Accessed from: https://www.linkedin.com/pulse/20140324055730-14091619-effectsof-web-3-0-in-the-new-digitalworld on 15 March 2020.

[3] Bhatt, R.K. & Kumar, A. (2014). Students' opinion on the use of social networking tools by libraries: A case study of Jawaharlal Nehru University, New Delhi. The Electronic Library, 32 (5), 594–602.

[4] Boyd, M.D. & Ellison, N.B. (2007). Social networking sites: definition, history and scholarship. Journal of Computer-Mediated Communication, 13(1), 210–230.

[5] Buabeng, C. (2012). Factors influencing teachers' adoption and integration of information and communication technology into teaching: A review of the literature. International Journal of Education and Development using Information and Communication Technology, 8(1), 136–155.

[6] Connell, R.S. (2009). Academic libraries, Facebook and Myspace, and student outreach: a survey of student opinion. Portal: Libraries and the Academy, 9(1), 25–36.

[7] Downes, S. (2005). "e-Learning 2.0, eLearn Magazine: Education and Technology in Perspective", available at: www.elearning.org/subpage.cfm?section1/4articles&article1/4 accessed on 25 April, 2020.

[8] Dul, J., Ceylan, C. & Jaspers, F. (2011). Knowledge workers' creativity and the role of the physical work environment. Human Resource Management, 50(6), 715–734.

[9] Evans, D. (2011) "The Internet of Things: How the Next Evolution of the Internet Is Changing Everything, white paper, Cisco Internet Business Solutions Group (IBSG)", Retrieved from: http://www.cisco.com/Web/about/ac79/docs/innov/IoT_IBSG_0411FINAL.pdf accessed as on 25 April 2020

[10] Feng, W.E.I. (2009). Research on the application of cloud computing on the internet, computer knowledge and technology. (1), 583–584, 670.

[11] Garoufallou, E. & Charitopoulou, V. (2011). Web 2.0 in library and information science education: the Greek case. New Library World, 113(4), 202–217.

[12] Geoghegan, M.W. & Klass, D. (2005) Podcast Solutions: The Complete Guide to Podcasting, Friends of ED, Berkley, CA. Glass, B. (2008), "Using Web 2.0 technologies to develop a sense of community for emerging LIS professionals", Proceedings of Meeting: 150. Continuing Professional Development and Workplace Learning, World Library and Information Congress: 74th IFLA General Conference and Council, 10-14 August 2008, Quebec, Canada, available at: http://archive.ifla.org/IV/ifla74/papers/150-Glass-en.pdf (accessed 9 November, 2019).

[13] Greenhow, C. (2009). Learning, teaching, and scholarship in a digital age: Web 2.0 and classroom research: what path should we take now?. Educational Researcher, 38(4), 246–559.

[14] Guntram, G. (2007) "Open educational practices and resources, OLCOS Roadmap 2012, Salzburg Research EduMedia Group, Salzburg," available at: www.olcos.org/cms/upload/docs/olcos_roadmap.pdf, accessed on 9 October, 2020.

[15] Hislop, D. (2013). Knowledge management in organizations: A critical introduction, 3rd Ed, UK, Oxford University Press.

[16] Hussain, F. (2012) E-learning 3.0 = E-learning 2.0 + WEB 3.0? IADIS International Conference on Cognition and Exploratory Learning in Digital Age (CELDA 2012).

[17] Joinson, A.N. (2008) "Looking at, looking up or keeping up with people? Motives and use of Facebook", Proceedings of the 26th Annual SIGCHI Conference on Human Factors in Computing Systems, ACM, Florence, Italy, 1027–1036.
[18] Kanagavel, R. & Velayutham, C. (2010). Impact of social networking on college students: a comparative study in India and The Netherlands. International Journal of Virtual Communities and Social Networking, 2(3), 55–67.
[19] Kelly, B. (2008). Library 2.0 and information literacy: the tools. In Godwin, P. & Parker, J., (Eds), Literacy Meets Library 2.0, Facet Publishing, London, Kennedy, G., Dalgarno, B., Gray, K. Judd, T., Waycott, J. Bennett, S., Maton, K. Krause, K. and Bishop, A. (2007). The next generation are not big users of Web 2.0 technologies: preliminary findings. Proceedings Asci lite Singapore 2007. Available at, http://ascilite.org.au/conferences/singapore07/procs/ken nedy.pdf (accessed 25 October, 2020).
[20] Lampe, C., Ellison, N. & Steinfeld, C. (2006). "A Facebook in the crowd: social searching vs social browsing", ACM Special Interest Group on Computer-supported Cooperative Work, ACM Press, Banff.
[21] Lee, H. & Choi, B. (2003). Knowledge management enablers, processes, and organizational performance: an integrative view and empirical examination. Journal of Management Information Systems, 20(1), 179–228.
[22] Lee, J. & Fink, D. (2013). Knowledge mapping: Encouragements and impediments to adoption. Journal of Knowledge Management, 17(1), 16–28.
[23] Liao, S.H., Chang, W.J., Hu, D.C. & Yueh, Y.L. (2012). Relationships among organizational culture, knowledge acquisition, organizational learning, and organizational innovation in Taiwan's banking and insurance industries. International Journal of Human Resource Management, 23(1), 52–70.
[24] Liebowitz, J. (2005). Linking social network analysis with the analytic hierarchy process for knowledge mapping in organizations. Journal of Knowledge Management, 9(1), 76–86.
[25] Luan, J. & Serban, A.M. (2002). Overview of Knowledge Management New direction for institutional research". Quarterly Review of Economic and Finance, 113, 605–631.
[26] Maness, J.M. (2006). Library 2.0 theory: Web 2.0 and its implications for libraries. Web ology, 2, available at www.webology.ir/2006v3n2.html?g¼link.webology.ir/ accessed as on 25 April 2020.
[27] Martensson, M. (2000). A critical review of knowledge management as a management tool. Journal of Knowledge Management, 4(3), 204–216.
[28] Morris, R.D. (2011). Web 3.0: implications for online learning. Tech Trends, 55(1), 42–46.
[29] National School Boards Association (2007), "Creating and connecting: research and guidelines on online social – and educational – networking", Available at: http://nsba.org/site/docs/41400/41340.pdf accessed as on 2 September 2020.
[30] Nonaka, I. (1991). The knowledge-creating company. Harvard Business Review, 69(6), (Nov.-Dec.), 96–104.
[31] Nonaka, I. (1994). A dynamic theory of organizational knowledge creation. Organization Science, 5(1), 14–37.
[32] Nonaka, I. & Takeuchi, H. (1995). The knowledge-creating company, Oxford University Press.
[33] Pan, S. & Scarbrough, H. (1999). Knowledge management in practice: An exploratory case study. Technology Analysis and Strategic Management, 11(3), 359–374.
[34] Peluchette, J. & Karl, K. (2010). Examining students' intended image on Facebook: 'what were they thinking?!. Journal of Education for Business, 85(1), 30–37.
[35] Prabhu, D. (2016). Application of web 2.0 and web 3.0: an overview. International Journal of Research in library science, 2(1), 54–62.

[36] Rai, K. (2011). Knowledge management and organizational culture: A theoretical integrative framework. Journal of Knowledge Management, 15(2), 779–801.
[37] Redecker, C. (2009). Review of Learning 2.0 practices: study on the impact of web innovations on education and training in Europe. JRC Scientific and Technical Reports, 23664, 1–122.
[38] Rehman, A.U. & Shafique, F. (2011) "Use of Web 2.0 and its implications for libraries: perceptions of information professionals in Pakistan", Library Philosophy and Practice, Available at: www.webpages.uidaho.edu/,mbolin/rehman-shafique.pdf accessed as on 5 October, 2020).
[39] Rogers, C.R. (2009). Social media, libraries and web 2.0. How American libraries are using new tools for public relations and to attract new users. In German Library Association annual conference: Deutscher Bibliothekartag, Erfurt, Germany.
[40] Rosen, C. (2007). Virtual friendship and the new narcissism. The New Atlantis, 17, 15–31.
[41] Santos, M. (2015). What is web 3.0 and why it is so important for business, Available at https://aquare.la/articles/2015/03/23/web-3-0-important-business/ accessed as on 28 September 2020.
[42] Serenko, A., Bontis, N., Booker, L., Sadeddin, K. & Hardie, T. (2010). A scientometric analysis of knowledge management and intellectual capital academic literature (1994–2008). Journal of Knowledge Management, 14(1), 3–23.
[43] Spender, J.C. (2008). Organizational learning and knowledge management: whence and wither?. Management Learning, 39(2), 158–176.
[44] Strati, A. (2007). Sensible knowledge and practice-based learning. Management Learning, 38 (1), 61–77.
[45] Virkus, S. & Bamigbola, A. (2011). Students' conceptions and experiences of Web 2.0 tools. New Library World, 112(12), 479–489.
[46] Wolfram, C. (2010) Q&A: Conrad Wolfram on communicating with apps in web 3.0, IT Pro, 17 March, Available at: https://www.itpro.co.uk/621535/qa-conrad-wolfram-oncommunicating-with-apps-in-web-30 accessed as on 28 September 2020.
[47] Yakhlef, A. (2010). The corporeality of practice-based learning. Organization Studies, 31(4), 409–430.
[48] Zack, M.H. (2003). Rethinking the knowledge-based organization. Sloan Management Review, 44(4), 67–71.

Vineet Chouhan, Shubham Goswami, Pranav Saraswat

# Knowledge Management Vis-à-Vis Adoption of Mobile Health Record Application for COVID Pandemic

**Abstract:** The use of electronic records has been widely accepted as an effective means of enhancing healthcare delivery. Despite advances in medical technology, healthcare continues to present worldwide difficulties of affordability, complexity, and access [1, 2]. Healthcare companies have issues regarding processes and information incorporation. Information is fragmented and often contradictory. Knowledge management can allow healthcare providers to access and exchange information about patients in present COVID situation. Healthcare systems are increasingly adapting to growing clinical demands that can be enhanced with the improvement of knowledge of the patients and using knowledge-based technology. Knowledge-based technologies can successfully help pandemic agencies by encouraging universal information sharing and rapid real-time monitoring. Acceptance of consumers is one of the core elements in the growth of information management systems (IMS). Research suggests a model of the variables that affect the contribution of KMS repositories. The research utilizes the unified principle of adoption and usage of technology (UTAUT) to define variables that may have an effect on the decision to use mobile health records (mHRA) to make a contribution to information.

**Keywords:** Mobile-Based Health Record Application (mHRA), Knowledge Management, COVID Pandemic

## 1 Introduction

Over the last decades, healthcare sectors have experienced tremendous changes in most of the countries in the world due to a rapid advancement in information and communication technology (ICT). Despite advances in medical technology, healthcare continues to present worldwide difficulties of affordability, complexity, and access [1, 2]. Healthcare companies have issues regarding processes and information incorporation. Information is fragmented and often contradictory. E-health needs improved knowledge management systems (KMS) to provide access to and interpret enterprise-wide knowledge for market intelligence. In comparison, information is not utilized for a competitive benefit, since the consistency of consumer choices is directly linked to the quality and availability of knowledge. These multiple knowledge bases cause inedibility incompatibility of format and limited ease of usage.

https://doi.org/10.1515/9783110722789-003

Moreover, number of KMS suffers from non or under use mainly because of a lack of contributions to the system. That is why an understanding of the drivers for adoption of KMS is a priority in knowledge management (KM) research.

In healthcare, KM is being developed mainly in the domain of electronic health record management and the amount of knowledge processing requirement in today's medicine is huge. In present pandemic time, IT tools that facilitate the capture and distribution of clinical information are becoming vital. Mobile application can support the creation; storage; retrieval; transfer and application of knowledge include data management and learning tools. These KM technologies are important components of KMS and have become central to healthcare [3, 4]. The amount of knowledge, created by all of healthcare partners, is tremendous. Information produced by one partner is of the utmost importance to all others for the delivery of quality treatment. Latest research shows that mHealth (mobile health) may be a blessing brought on by ICT and is undoubtedly one of the most popular programs with significant impacts on the growth of KM in the healthcare field.

In this respect, mHealth technology promotes KM and also has the ability to enhance public health systems in developing economies like India. Acceptance of mobile devices in the health record may create awareness as a platform that signals quality health services and gives the public health service a good image [5]. Technology allows the creation and capture of information. The architecture promotes the codification of information and the diffusion of knowledge to standardize the level of patient care. The primary goal for EHR systems and related technologies should be to facilitate KM for improving individual and community health. Moreover, it has been argued that EHR system is a must have tool for public health officials to improve treatment, care, and prevention COVID and its diagnosis.

Information management can help mitigate medical mistakes and thus their costs by giving decision-making assistance to clinicians. Healthcare is a dynamic area, and collaboration between various providers of healthcare is essential for the delivery of quality care. Hence, cooperative diagnosis can be achieved by the healthcare actors via the implementation of KMS. Past studies showed strong evidence that KM implementation in healthcare is encouraging, that can impact decision-making. It can be said that implementing KM in healthcare, such as advances in healthcare communication technology, mobile technology can be the way forward to improve the quality of care for patients, which is the ultimate goal of healthcare [3, 4].

## 2 Review of literature

Software and gadgets allow us to log in our operation, provide guidance and boost our efficiency in certain instances [6]. Free wellness is an affordable and flexible way for us to be vigilant at self-managing our well-being, from app push reminders to wearables that provide sensory feedback [7]. Device abundance in wellbeing is now available in devices and wearables to better tackle safety, wellness and numerous therapies issues. These Even, they give updates and nudges to hold us on board. The usage of different technology technologies and mHealth apps has often addressed several other (high) risk behaviors. In India, although several applications focused on the smartphone are being created. The national health portal identified 74 apps that can be used for health information. Still, newly created 'My Health Record' and 'Aarogya Setu' apps are being established by the Ministry of Health during and after COVID. It may allow for the early diagnosis of numerous medical disorders, including certain types of diseases that can with COVID, harm the human lives. When the proliferation of smartphones and portable apps grows, their usage further encourages a more patient-centric attitude. Advanced digital health infrastructure access is transforming medical treatment by providing new approaches to track and treat illnesses and make care more useful and comprehensive [8].

mHealth application is seen by those in need of healthcare like India as being a simple and inexpensive way to improve their access to health services. The increasing smartphone user in the county also provides an opportunity to develop a mobile-based health record system to improve healthcare. It is also evident from the past studies that mHRA have capabilities transforms healthcare from simple monitoring self-healthcare to supervised and assisted healthcare services. Despite the possible advantages of mHRA systems, recent findings suggest that approving such proposals appears to be a challenge. Ahmed et al. [9] said that many people, while few have used mHRA, were conscious of and considered to be potentially useful. Just 5% of participants in the study used mobile facilities on their phones, and 73% of respondents knew about the potential but were not able to do so. The research Yi et al. [10] has found influences that affect mHealth services acceptance such as perceived ease of use, program consistency, subjective normativity, perceived value, device performance, and technical anxietary. In Bangladesh, Akter et al. [11] have studied the standard of service mHealth adoption. It was found to affect mHealth systems both directly and indirectly. All the above studies examine mHRA acceptance but provided diverse outcomes regarding adoption. Such studies indicate however clearly that most individuals in India are aware of this technology and are involved in it. Despite the potential advantages and the various mHealth initiatives in place, recent studies show that the adoption of this service remains insignificant.

The report on the implementation and usage of mHRA to support KMS from end-user perspective has been done little work. To date, most studies have continued to concentrate on lifestyle, user experience, and security acceptance of mHRA [3, 12–14], resistance and technology anxiety, and attitudes. Moreover, prior readings have primarily engrossed on medical specialized views for implementation [15]. Few studies have focused on the end-user approval of Health application to contribute to KMS. However, this study primarily emphasizes on identifying barriers in adopting mHRA by using a human-centric perspective and applying socio-technical characteristics of facilitating condition and social influence, which have not been spoke about the implementation of mHRA facilities.

As limited studies have been conducted on health information system implementation in emerging republics, makes the present study outline more relevant and necessary. With respect given the fiction review, the present research attempt to fill above gaps by examining the recognition issues of mHRA in a developing country like India Acceptance by consumers is one of the main elements for the advancement of information management systems. The research suggests a model of the variables that affect the contribution of KMS repositories. Specifically, in the present study, a UTAUT-established prototypical is cast-off to consider the behavioral purpose of end-users in India to agree to take and use mHRA for knowledge contribution.

# 3 Theoretical framework

The Unified Theory of Acceptance and Use of Technology (UTAUT) is used, as the speculative outline to empirically test the factors that impact the adoption of mHRA developing country context [16]. The UTAUT is the most popular user approved and behaviors analysis model, as it is derived from theory of planned behavior (TPB); theory of reasoned action (TRA); technology acceptance model (TAM); innovation diffusion theory (IDT); PC utilization model (PUM), and the social cognitive theory (SCT). The model is aimed to attain a unified opinion of consumer recognition, is and appealed to predict 70% of the modification in purpose to use new technologies [17].

Using a review of prior studies of UTAUT in healthcare [18–20], the most used structures in this analysis are the latest science. The computational analysis model includes the systems commitment expectativeness (EE), Performance Expectancy (PE), Facilitating Condition (FC), social impact (SI), and use behavior (UB). As UTAUT is a most robust framework for predicting technology acceptance with good predictability. However, in particular in developed nations, its sufficiency in forecasting technological acceptance in healthcare is uncertain [21]. Some authors used UTAUT variables with theoretical models for determining antecedents in healthcare [22–24]. Because mHRA adoption is highly context-specific [25], it is important to examine the opinions of machine users.

In order to increase predictability of UTAUT in a healthcare environment the addition of domain-specific determinants can also be considered [26, 27]. Technology adoption in developing countries like India is influenced by background specific societal values and confidence concerns related to obstacles, such as the digital divide. In addition, in order to increase its predictive capacity, context-specific variables are applied to the simple UTAUT model. An updated UTAUT is then used to suit further into the study context. Constructs of self-efficiency and trust dimensions have been added to the model. This is due to the perceived lack of experience and its associated trust issues related to data privacy issues with a mobile application in the Indian context. The following section explains each construct with hypotheses developed to investigate the relationships. It is defined as a degree to which an individual believes that using the system will help to gain healthcare experience.

## 3.1 Performance expectancy (PE)

Venkatesh et al. [16, 17, 21] have found that PE is the strongest determinant of a user's behavioral intention (BI) and was identified as convenience expectancy, outcome expectancy, and time expectancy. Studies by Pai and Huang [28], Sun et al. [29] and Carlsson [30] also confirmed the significant positive influence of PE on the intention to use mHealth services.

The above discussion led this study to posit the following hypothesis:

**H1. PE has a positive impact on the intention to use mHRA**

## 3.2 Effort expectancy (EE)

It is defined as the ease with which the mHRA is used. Past experiments have shown that EE has a major effect on the desire of consumers to use mHRA [29, 30]. The use of mHealth tracking devices through smartphones has been described as a major motivator [31]. Therefore, it is assumed:

**H2. EE has a positive impact on the intention to use mHRA**

## 3.3 Social influence (SI)

The extent to which a person understands the value of peers' viewpoints and can use mobile medical record applications is defined. Work has shown that SI significantly influences the desire of consumers to use technology [29, 31]. Wills et al. [31, 32] have reported that the behavioral aim for mobile digital wellbeing is an important association between the IS and consumers. The following theory was raised in line with the aforementioned discussion:

**H3. SI has a positive impact on the intention to use mHRA**

## 3.4 Self-efficiency (SE)

It is the belief of performing a particular behavior. Self-efficacy perceptions have been found to influence decisions about what actions to undertake [34]. Educational researchers have employed self-efficacy instruments in various disciplines and academic settings [35]. Studies also measure self-efficacy specific to the online environment [34, 36, 37]. Cho and Castañeda [38] studied personal dedication with a smartphone application to learn Spanish. Few types of research also study found the Motivational and emotional commitment to learning Spanish with a mobile application. Few studies also found self-efficacy positive impact on the intention to adopt mHealth apps [39, 40]. The following theory was raised according to the discussion:

**H4. SE has a positive impact on the intention to use mHRA**

## 3.5 Trust (TR)

Views on security and confidentiality have received interest in recent literature on online transactions. The perception of risk to safety and threat of personal data loss may differ with a change in demographic of consumer groups. This negative emotional response, the intention of adoption is diminished by anxiety or frustration with technology. Older adults are also reported to have significant problems of confidence [41]. This analysis thus expands the fundamental UTAUT with an additional confidence variable to increase the prediction power. The following theories have also been suggested:

**H5. TR harms the intention to use mHRA**

## 3.6 Facilitating conditions (FC)

It can be described as to what degree a person assumes that the usage of mHRA is enabled by an organizational and technological infrastructure. Conditions of facilitation require service scale products, protection and secrecy, technical difficulties. The direct effect of FC technology help on the behavioral purpose and technical use is supported by previous studies [10, 42]. The Boontarig et al. [31] research has demonstrated that FC has had a positive effect on the behavior. The argument above lead to these hypotheses:

**H6. FC has a positive impact on the intention to use mHRA**

**H7. FC has a positive impact on the actual use of mHRA**

In the literature on technical implementation, the relation between behavioral Intention (BI) and actual action (UB) is well established and notes that BI is a reliable predictor for actual use. [17, 21, 41]. The present study also tests this relation, and the following hypotheses were proposed:

**H8. The BI has a positive impact on the actual use of mHRA**

Individual stronger orientation toward knowledge is more associated with the notion of usage. We accept the positive relationship existing between intention to use mHRA systems and knowledge sharing behavior. To test this relation, the following hypothesis was proposed:

**H9. The usage behavior has a positive impact on perception toward contribution to KMS Figure 1 present the Research Model and Table 1 presents the summary of construst with scale items as under:**

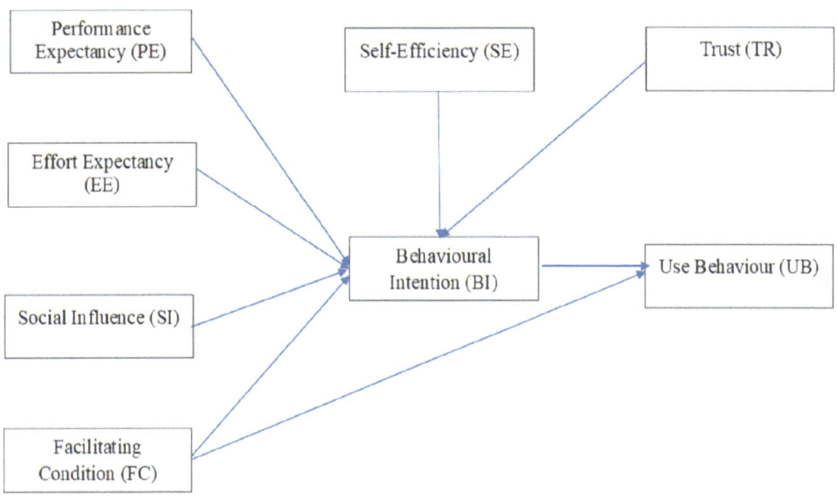


**Figure 1:** Research model.

**Table 1:** Summary of constructs with scale items.

| Constructs | Scale items |
|---|---|
| Performance expectancy (PE) | PE1. I find mHealth applications useful in my daily life. |
| | PE2. Using mHealth applications help to accomplish things more quickly.<br>PE3. mHealth applications increase my productivity. |
| Effort expectancy (EE) | EE1. Learning how to use mHealth applications is easy.<br>EE2. Interaction with mHealth applications is clear and understandable. |
| | EE3. I find mHealth applications service easy to use.<br>EE4. It is easy for me to become skillful in using mHealth applications. |

**Table 1** (continued)

| Constructs | Scale items |
|---|---|
| Social influence (SI) | SI1. People who are important to me suggest that I should start utilizing mHealth apps. |
| | SI2. People who affect my actions believe I can use mHealth applications.<br>SI3. People whose views I respect tend to use mHealth applications. |
| Facilitating condition (FC) | FC1. I have the tools to allow use of mHealth apps. |
| | FC2. I have the knowledge necessary to use mHealth applications.<br>FC3. mHealth applications are compatible with other technologies I use. |
| Self-efficiency (SE) | SE1. I can use mealth application without the guidance of anyone. |
| | SE3. I believe that most issues with the use of mHealth apps are dealt with effectively.<br>SE3. I am not nervous when I face difficulty with the system. |
| Trust (TR) | TR1. mHealth applications have adequate security features. |
| | TR2. I feel safe in my transaction with a mHealth application.<br>TR3. I trust recommendations provided by mHealth application.<br>TR4. I trust mHealth application for data privacy. |
| Behavioral intention (BI) | BI1. I expect to use mHealth applications. |
| | BI2. I will continuously trying to use mHealth applications in my daily life.<br>BI3. I proposed to use mHealth applications service frequently. |
| Use behavior (UB) | UB1. Using mHealth applications is a pleasant experience. |
| | UB2. I use mHealth applications currently.<br>UB3. I spend a lot of time on mHealth applications. |
| Contribution to KMS (KMS) | KM1. I am willing to share my health information to knowledge base. |
| | KM2. I am ready to learn the skills required for knowledge sharing.<br>KM3. I like to contribute my knowledge for development of integrated KMS. |

# 4 Methods

All the predictors for the modified model (EE, PE, SI, FC, SE, TR, BI and Use Behaviour (UB)) are measured using scale items adapted from similar studies in healthcare.

In the first section, population information is included in the questionnaire used in the survey and in part B, questioning for the various buildings mentioned in the sample sequence. The pilot study of 25 respondents was performed to test the suitability of the questionnaire. The pilot analysis is used to render contextual changes.

The census of research consisted of smartphone users in select major cities of India, representing the urban population of the country. Due to their age group (≥60), their technological skills and their mobile / internet use experience they were chosen. Responses from survey items measuring theoretical constructs captured via a 5-point Likert scale with reactions ranging from "strongly disagree" (1) to "strongly agree" (5).

# 5 Result

The research circulated 3,000 **questionnaires online and 2,140 of them were analyzed**. In terms of demographic profiles, 56% of males and 44% of people who participated in the study belong to various stages of age. The majority of respondents belonged to the service sector (55%) and aged between 30 and 50 years. In terms of access, the majority of respondents owned a smartphone, with 65% of them are using it for five years. A total of 58% of respondents have experiences in using mHRA, 42% are not using but willing to adopt the mobile smartphone-based application. The respondents profile is presented in table 2.

**Table 2:** Respondents profile.

| Category | Percentage |
|---|---|
| Gender | |
| Male | 56% |
| Female | 44% |
| Age | |
| 20–30 years | 23% |
| 30–40 years | 38% |
| 40–50 years | 39% |
| Profession | |
| Service | 55% |
| Business | 32% |
| Student | 13% |
| Experience in using the smartphone | |
| 0–5 years | 65% |
| 5–10 years | 23% |
| >10 years | 12% |
| Experience in using mHRA | |
| Yes | 58% |
| No | 42% |

The first phase of analysis, the validity of the measurement model is tested. Item Loadings of all the factors are higher than the threshold value of 0.5. This implies that these variables converge on a Latent Variable common point. It proves the integrity of the house. Composite reliability (CR) of all the observed variables is higher than 0.7 indicate excellent safety of the factor structure. The internal reliability was evaluated considering Cronbach's alpha. The Alpha Cronbach values of the factors exceeded 0.7, so that the variable can be accurately evaluated further. In order to assess the discriminating validity of each house, the average vector extraction (AVE). The model fits indices used for measuring the data's performance (Table 3).

**Table 3:** Measurement model.

| Constructs | Items | Loading | AVE | CR | Cronbach alpha |
|---|---|---|---|---|---|
| Performance expectancy (PE) | PE1 | 0.702 | 0.873 | 0.817 | 0.897 |
| | PE2 | 0.826 | | | |
| | PE3 | 0.880 | | | |
| Effort expectancy (EE) | EE1 | 0.863 | 0.516 | 0.873 | 0.893 |
| | EE2 | 0.757 | | | |
| | EE3 | 0.724 | | | |
| | EE4 | 0.858 | | | |
| Social influence (SI) | SI1 | 0.897 | 0.634 | 0.783 | 0.786 |
| | SI2 | 0.812 | | | |
| | SI3 | 0.730 | | | |
| Facilitating condition (FC) | FC1 | 0.825 | 0.636 | 0.927 | 0.792 |
| | FC2 | 0.714 | | | |
| | FC3 | 0.820 | | | |
| Self-efficiency (SE) | SE1 | 0.743 | 0.685 | 0.813 | 0.816 |
| | SE3 | 0.734 | | | |
| | SE3 | 0.967 | | | |
| Trust (TR) | TR1 | 0.850 | 0.772 | 0.842 | 0.845 |
| | TR2 | 0.682 | | | |
| | TR3 | 0.750 | | | |
| | TR4 | 0.751 | | | |
| Behavioral intention (BI) | BI1 | 0.901 | 0.678 | 0.741 | 0.833 |
| | BI2 | 0.595 | | | |
| | BI3 | 0.732 | | | |
| Use behavior (UB) | UB1 | 0.622 | 0.743 | 0.811 | 0.767 |
| | UB2 | 0.669 | | | |
| | UB3 | 0.575 | | | |
| Contribution toward KMS (KMS) | KM1 | 0.743 | 0.764 | 0.832 | 0.888 |
| | KM2 | 0.790 | | | |
| | KM3 | 0.696 | | | |

In order to analyze the relationship between the structures in the research model, the structural model was established. The path coefficients were also determined to assess the effect of each element on the deciding variable (as per Table 4) at this second step of research. The path coefficients (parameter estimates) were used to assess the magnitude and direction of relationships and also to test the research hypotheses. The structural model was estimated by AMOS 22 to represents relationships among observed and unobserved variables using path diagrams. The greater value of $R^2$ revealed better model quality. The results revealed that all the hypotheses are significant. The results show that the relationships between PE and BI (B = 0.397, $p < 0.05$); EE and BI (B = 0.252, $p < 0.05$); SI and BI (B = 0.235, $p < 0.05$); SE and BI (B = 0.412, $p < 0.05$); FC and BI (B = 0.213, $p < 0.05$); FC and UB (B = 0.374, $p < 0.05$); and BI and UB (B = 0.381, $p < 0.05$), and UB and KMS (B = 0.256, $p < 0.05$) is positive and significant. Furthermore, the path coefficient between TR and BI (B = −0.078, $p > 0.05$) is not significant. Therefore, the majority of hypotheses formulated to test the relationship are significant.

**Table 4:** Result of the structural model and model fit indices.

| Path | | | Estimate | C.R. | P | Hypothesis |
|---|---|---|---|---|---|---|
| PE | → | BI | 0.297 | 5.752 | *** | H1 (support) |
| EE | → | BI | 0.252 | 6.069 | *** | H2(support) |
| SI | → | BI | 0.358 | 5.296 | 0.022 | H3(support) |
| SE | → | BI | 0.412 | 4.845 | *** | H4(support) |
| TR | → | BI | −0.078 | 3.982 | 0.113 | H5(not supported) |
| FC | → | BI | 0.213 | 6.769 | *** | H6(supported) |
| FC | → | UB | 0.374 | 5.877 | *** | H7(supported) |
| BI | → | UB | 0.381 | 5.983 | *** | H8(supported) |
| UB | → | KMS | 0.256 | 4.615 | 0.031 | H9(supported) |

**Model fit indices**

$\chi^2$/df = 2.614; GFI = 0.922; CFI = 0.932; NFI = 0.935; AGFI = 0.931 and RMSEA = 0.042.

***Significant at $p < 0.001$ level.

# 6 Conclusion and discussion

The research analyzed causal factors for use and implementation of mHRA by evaluating a UTAUT model. End-user acceptance is essential in order for any smartphone application to be successful. Given the strong context-specific use and usage of these programs, faith and auto-efficiency variables are also used in the study of acceptance. The review of the literature shows that few acceptance trials in developed countries have been carried out, in particular using UTAUT. In developed countries, there is little information about UTAUT's adequacy in mHealth adoption. This thesis was based on a sound technological recognition paradigm and testing methods. The findings of our research clearly indicate that patients should provide patient reports that they find important for better safety and healthcare quality.

The sample was comprised exclusively from smartphone users across major urban locations in India. This research uses the UTAUT model to determine the users' BI to use Mobile-based health applications. The findings have given a snapshot into UTAUT buildings such as PE, EE, SI, SE, TR, and FC which affect the adoption. The research confirms the suggested conclusions on buildings in operation. Furthermore, the behavioral response was measured using a multidimensional method. The findings indicate the import variables that are helpful to boost their efficiency in implementing mHRA. For a productivity effect, the factors analyzed are critical. Seven of our eight theories for study have been endorsed with high statistical significance.

The findings of this work are consistent with previous work and BI studies in embracing and using mHealth. The findings of previous research in the use of the UTAUT implementation model in e-health applications are consistent. PE and EE have major impacts on the application of health information systems (HIS) according to Phichitchaisopa and Naenna [43]. The main effects of PE and EE on adoption of electro-medical documentation (EMR) have been confirmed by Hennington et al [44]. Holtz and Krein [45] also found a positive relation between PE and SI in the adoption of EMR systems. Wills et al. [33] indicated that SI is more significant in an environment where the user has inadequate experience of online services. In India, cell phones are also a new phenomenon for others than telephony calls, such as health advice.

Such findings suggest that the architecture of EE allows to embrace and use a mHRA. It also says that an easy-to-use interface is important. To use mHRA, it should be efficient and user-friendly in conjunction with the desired workloads. The findings of our analysis are similar to the findings from UTAUT [16] with less direct effect. It can be explained with the low maturity of mobile applications in India. FC like training, online technical support, and IT infrastructure needed to accommodate such systems in developing countries. As a consequence, it is possible that such programs will fail until the program has fulfilled the infrastructural needs. Social effect (SI) relies heavily on people around the consumer in adoption. It can be concluded that most of them were found more likely to install the application, if it became mandatory and if other in the group is using it.

In a multidimensional system the existence of a heavy influence by SE and EE was obtained. In this context, self-esteem, and the desire to use the program seem to play an important role as a way to affect the success of improved use of the mHRA. The construction of SE often leads greatly to the approval of the enterprise. The paper better understand the adoption of KMS with the help of the application of a model, UTAUT, adapted to the specific context of KM. Findings thus validate the use of the UTAUT model in the specific context of KMS. Using an empirical study, researcher validated a part of UTAUT, relating to intentions to contribute to KMS.

# 7 Implications and limitations of the study

Measurement of the behaviors and the intention of consumers to use the program when the purpose is to seek an explanation of the actual use of the system. The results show that to increase the use of mHRA to support clinical practice; it is essential to focus on the usefulness, user-friendliness to improve the SE of the application – most of the studies in a healthcare measure the effectiveness of the systems. But with the current research on mHRA acceptance, we focus on the actual behavior of end-users. Results are useful for decision makers, healthcare administrators and clinicians in developing economies. The findings of this research add greatly to literature potentially of healthcare application. KM insights are given for the deployment of mHealth technologies for frontline health professionals in developing economies. The research used frequently applied technology acceptance model for studying user acceptance of the technology. Still, only a few studies have investigated technology acceptance from an end-user perspective for more applications. For a developed world or in emerging countries with little focus on implementing MHRA, most recent studies follow the UTAUT pattern.

KM can allow healthcare providers to access and exchange information about patients in present COVID situation. Healthcare systems are increasingly adapting to growing clinical demands that can be enhanced with the improvement of knowledge of the patients and using knowledge-based technology. Knowledge-based technologies can successfully help pandemic agencies by encouraging universal information sharing and rapid real-time monitoring. The study showed that the behavioral intent to use the mHRA has considerable effect on success anticipation, commitment expectations, social control and productive self-management, and also findings indicate that the usage of the software does not have a substantial association between the use of the confidence and the consumer behavior. Findings provide service providers and decision makers with useful knowledge in designing a plan to incorporate this technology to motivate users for knowledge creation and sharing. In this research, policy makers and decision makers plan to introduce mHealth apps in this country also make significant contributions in developing a comprehensive KM framework. The Indian government is setting the basis to carry out the national health plan. The variation in outcomes is based on the study context and the dependent variables chosen and previous experiments are based on them. In this analysis, end-user opinions and the purpose to use the program are known. That would make the decision-making process more educated and improve the potential for contribution towards knowledge creation, storage and sharing.

Empirical studies provide realistic advice in developed countries like India on a successful mHRA adoption and knowledge sharing attitude. For an expanded consumer awareness Mobile-based Health record application services, developers and providers of application can understand the challenges regarding the design of mHRA. This study also gives importance that in order to enhance UTAUT's explanatory mode power in the context of developing countries, researchers should use context-specific dimensions to modify it. The survey can provide developers with a framework in which BI and real use focusing on age demographics can be influenced. This study could be modified easily to help other developing countries implement similar health-based applications due to a generic approach.

Another result deserves particular attention, which is the link between usage behavior and contribution to KMS. The theoretical hypothesis supports a relationship between the two variables. This revealed the fact that experienced individuals integrate the sharing of knowledge and can develop positive attitude for contributing toward KMS. Implementation of KM in healthcare can provide opportunities in providing the right knowledge at the right time, i.e., at the point of decision making by implementing KM in healthcare is paramount. Appropriate tool like mobile-based systems and user-friendly interface is a requirement as it can significantly improve the quality and safety of care provided for patients at both hospital and home settings. At present evidence-based medical practice is critical for developing countries. However, this can be applied successfully if and only if the right system is in place for KM, and thus, it should be given due attention. It is important to allow use of the available developments in mobile devices and decision support systems, electronic health record systems.

Future research should concern the importance of testing a complete model that covers cognitive, affective, and behavioral responses. Limitation of the study includes reduced sample size belong to only urban cities in the country. Thus, it also raises concern about the generalizability of the findings. Future studies can expand geographic coverage and include more demographic variables in studies. The responses also suffer from the bounded rationality since the users may be in the early stages of adoption of mobile-based health record system. Further, the actual usage is measured with self-reported actual behavior, which is not a direct measure that can be generated by system logs. Another extension of the present study could adopt both quantitative and qualitative data analysis approaches to uncover users' in-depth opinions on acceptance and usage issues.

## References

[1] Panesar, A. & Panesar, H. (2020). Artificial intelligence and machine learning in global healthcare. Handbook of Global Health, 1(1), 1–39.

[2] Barzotto, M., Corradini, C., Fai, F.M., Labory, S. & Tomlinson, P.R., Eds. (2020). Revitalising Lagging Regions: Smart Specialisation and Industry 4.0, Routledge.

[3] Shahmoradi, L., Safadari, R. & Jimma, W. (2017). Knowledge management implementation and the tools utilized in healthcare for evidence-based decision making: A systematic review. Ethiopian journal of health sciences, 275, 541–558.

[4] Jimma, W. (2017). Knowledge management implementation and the tools utilized in healthcare for evidence-based decision making. Ethiopian Journal of Health Sciences, 27,(5), 541–548.
[5] Fletcher-Brown, J., Carter, D., Pereira, V. & Chandwani, R. (2020). Mobile technology to give a resource-based knowledge management advantage to community health nurses in an emerging economies context. Journal of Knowledge Management, Vol. ahead-of-print No. ahead-of-print. doi: https://doi.org/10.1108/JKM-01-2020-0018.
[6] Schoonveld, E. (2020). The Price of Global Health: Drug Pricing Strategies to Balance Patient Access and the Funding of Innovation, Routledge.
[7] Noori, N., Hoppe, T. & De Jong, M. (2020). Classifying pathways for smart city development: Comparing design, governance and implementation in Amsterdam, Barcelona, Dubai, and Abu Dhabi. Sustainability, 12(10), 4030–4054.
[8] Wójtowicz, A. & Chmielewski, J. (2020). Analysis of factors improving accuracy of passive user identification with streams of face images for ubiquitous commerce. In International Conference on Human-Computer Interaction, Springer, Cham.
[9] Bloom, A.G., Iqbal, M., Lucas, H., Rasheed, S., Waldman, L. & Bhuiya, A. (2014). E-health and M-health in Bangladesh: Opportunities and Challenges.
[10] Yi, M.Y., Jackson, J.D., Park, J.S. & Probst, J.C. (2006). Understanding information technology acceptance by individual professionals: Toward an integrative view. Information and Management, 43,, 350–363.
[11] Akter, S., D'Ambra, J. & Ray, P. (2010). Service quality of mHealth platforms: Development and validation of a hierarchical model using PLS. Electron Mark, 20, 209–227.
[12] Lin, C.P. & Anol, B. (2008). learning online social support: An investigation of network information technology based on UTAUT. Cyberpsychology Behavior, 11, 268–272.
[13] Pai, F.Y. & Huang, K.I. (2011). Applying the technology acceptance model to the introduction of health-care information systems. Technological Forecasting and Social Change, 78, 650–660.
[14] Nuq, P. & Aubert, A.B. (2013). towards a better understanding of the intention to use eHealth services by medical professionals: The case of developing countries. International Journal of Healthcare Management, 64, 217–236.
[15] Carlsson, B. (2006). Internationalisation of innovation systems: A survey of the literature. Reserach Policy, 35, 56–67.
[16] Venkatesh, V., Morris, M.G., Davis, G.B. & Davis, F.D. (2003)User acceptance of information technology: Toward a united view. MIS Q, 27(3).
[17] Martín, H.S. & Herrero, Á. (2012). Influence of the user's psychological factors on the online purchase intention in rural tourism: Integrating innovativeness to the UTAUT framework. Tour Management, 33(2), 341–350.
[18] Cheng, D., Liu, G. & Qian, C. (2008). On determinants of user acceptance of internet banking: A theoretical framework and empirical study, 2008 IEEE Symposium on Advanced Management of Information for Globalised Enterprises AMIGE.
[19] Hoque, R. & Sorwar, G. (2017). Understanding factors influencing the adoption of mhealth by the elderly: An extension of the UTAUT model. International Journal of Medical Informatics, 101, 75–84.
[20] Venkatesh, V., Sykes, T.A. & Zhang, X. (2011). Just what the doctor ordered': A revised UTA for EMR system adoption and use by doctors, 2011, 44th Hawaii International Conference on System Sciences.
[21] Kim, S., et al. (2015). Analysis of the factors influencing health-care professionals' adoption of mobile electronic medical record (EMR) using the united theory of acceptance and use of

technology (UTAUT) in a tertiary hospital. BMC Medical Informatics and Decision Making, 16(12), 1–12.
[22] Weeger, A. & Gewald, H. (2015). Acceptance and use of electronic medical records: An exploratory study of hospital physicians' salient beliefs about HIT systems. Health System, 41, 64–81.
[23] Gagnon, M.P., Jose, A. & Emparanza, J. (2012). Using a modified technology acceptance model to evaluate healthcare professionals' adoption of a new telemonitoring system, in telemedicine and e-health. Mary Ann Liebert, Inc, 54–59.
[24] Andargoli, E., et al. (2017). Health information systems evaluation frameworks: A systematic review. International Journal of Medical Informatics, 9(7), 195–209.
[25] Succi, M.J. & Walter, Z.D. (1999). Theory of user acceptance of information technologies: A examination of health care professionals, Systems Sciences, 1999, HICSS-32. Proceedings of the 32nd Annual Hawaii International Conference on, IEEE.
[26] Ahlan, A.B.D. & Rahman Ahmad, B.I.E. (2014). User acceptance of health information technology HIT in developing countries: A conceptual model. Procedia Technology, 16,, 1287–1296.
[27] Pai, F.Y. & Huang, K.I. (2011). Applying the technology acceptance model to the introduction of health-care information systems. Technological Forecasting and Social Change, 78, 650–660.
[28] Sun, N., Wang, X., Guo, X. & Peng, Z. (2013). Understanding the acceptance of mobile health services: A comparison and integration of alternative models. Journal of Electronic Commerce Research, 14(2), 183–200.
[29] Carlsson, B. (2006). Internationalisation of innovation systems: A survey of the literature. Reserach Policy, 35, 56–67.
[30] Boontarig, W., Chutimaskul, W., Chongsuphajaisiddhi, V. & Papasratorn, B. (2012). Factors influencing the Thai elderly intention to use smartphone for e-Health services, IEEE Symposium on Humanities, Science and Engineering Research SHUSER, 479–483.
[31] Lu, J., Yao, J.E. & Yu, C.S. (2005). Personal innovativeness: Social influences and adoption of wireless Internet services via mobile technology. Journal of Strategic Information Systems, 14, 245–268.
[32] Wills, M.J., El-Gayar, O.F. & Bennett, D. (2008). examining health-care professionals' acceptance of electronic medical records using UTAUT. Information Systems, 9, 396–401.
[33] DeTure, M. (2004). Cognitive style and self-efficacy: Predicting student success in online distance education. American Journal of Distance Education, 18(1), 21–38.
[34] Hayashi, A., Chen, C., Ryan, T. & Wu, J. (2020). The role of social presence and moderating role of computer self-efficacy in predicting the continuance usage of e-learning systems. Journal of Information Systems Education, 15(2), 1–5.
[35] Artino, J.A.R. & McCoach, D.B. (2008). Development and initial validation of the online learning value and self-efficacy scale. Journal of Educational Computing Research, 383, 279–303.
[36] Tsai, M.J., Wang, C.Y. & Hsu, P.F. (2019). Developing the computer programming self-efficacy scale for computer literacy education. Journal of Educational Computing Research, 568, 1345–1360.
[37] Cho, M.H. & Castañeda, D.A. (2019). Motivational and affective engagement in learning Spanish with a mobile application. System, 81, 90–99.
[38] Balapour, A., Reychav, I., Sabherwal, R. & Azuri, J. (2019). Mobile technology identity and self-efficacy: Implications for the adoption of clinically supported mobile health apps. International Journal of Information Management, 49, 58–68.

[39] Reychav, I., Beeri, R., Balapour, A., Raban, D.R., Sabherwal, R. & Azuri, J. (2019). How reliable are self-assessments using mobile technology in health-care? The effects of technology identity and self-efficacy. Computers in Human Behavior, 91, 52–61.
[40] Hoque, R. & Sorwar, G. (2017). Understanding factors influencing the adoption of mHealth by the elderly: An extension of the UTAUT model. International Journal of Medical Informatics, 101, 75–84.
[41] Bhattacherjee, A. & Hikmet, N. (2008). Reconceptualising organisational support and its effect on information technology usage: Evidence from the health care sector. Journal of Computing and Information System, 48.
[42] Phichitchaisopa, N. & Naenna, T. (2013). Factors affecting the adoption of health-care information technology. EXCLI Journal, 12, 413–436.
[43] Hennington, A. & Janz, B.D. (2007). Information systems and health-care XVI: Physician adoption of electronic medical records: Applying the UTAUT model in a health-care context. Communications of the Association for Information Systems, 19(5).
[44] Holtz, B. & Krein, S. (2011). Understanding nurse perceptions of a newly implemented electronic medical record system. Journal of Technology in Human Services, 29, 247–262.

Navita Mahajan, Seema Garg, Oridoye Tomide Hakeem

# Diversified Landscape of Knowledge Management: Transformation in View of Organizational Sustainability

## Learning objectives:

- ***Knowledge and knowledge-based view of firms***
- ***Insight into knowledge management with diversified landscape for organizational sustainability***
- ***Knowledge transformation and web transformation in view of Web 1.0 to Web 3.0***

**Abstract:** The changing landscape of modern society, diversified aspects of global economy, knowledge-based models, and multidimensional theories of firms require global perspectives of innovative thoughts. Organizations are to be prepared for the future and scrutinized that what these businesses are doing best from the rest. The competitive environment of firms is changing swiftly, hence knowledge-based management and its transformational practices are crucial for sustainability and self-reliance of any businesses. The time has come for traditional sources of knowledge to get new paradigms, technological changes, highly advanced modern practices, intangible characters, and pertinent information, all gaining priority in a firm for long-term sustainability. The advanced platforms such as Web 3.0 processes for increased efficiencies are creating incentives for innovation, resulting in entrepreneurs seeking opportunities in changing dynamics and then adapting to these transformations.

To get more insight into the nature and content of knowledge, this chapter traces the transformational basis of knowledge management in terms of the emergence of the knowledge-based view of the firm, its changing dimensions, and a multifaceted review of received literature in which author has examined the knowledge models concepts through multiple subject matter lenses. Knowledge management has provided an umbrella for a range of management activities – all linked together by their central focus on knowledge within the organization. By focusing on some of the core elements of knowledge led theories of firm, changing aspects of knowledge management system, it has been found that those factors have been highlighted which influence firm's performance, ability of affirms to construct and design the landscape for effectiveness.

The chapter has covered knowledge-based view of the firms wherein the theoretical approaches from economies and organization theory in changing scenario have been discussed by focusing on theory of firm. Later on the knowledge variations in terms of technology, innovation, science, acquaintance description, explicit, tacit

https://doi.org/10.1515/9783110722789-004

interfaces along with paradigms of constructing scientific knowledge are depicted. Later on, the study links knowledge innovations' influence in view of new economic and social environment with advancement in Web 3.0 technology for competitiveness of organizations. The chapter analyzes different approaches to share, develop, and formalize knowledge for sustainable new approaches in an organization.

**Keywords:** Knowledge, Landscape, Models, Sustainability, KM, Web, Innovation

# 1 Introduction

Knowledge is the prime and supreme for most of the firm's competitive advantage. This includes both codified knowledge such as written documents and blueprints and tacit knowledge such as routines. Also, it has been revealed that firms that do a better job of protecting their unique knowledge from being misused, undergo any theft, observed, and imitated, will always have an advantage over those that don't have these characteristics. It can be argued that it is essential to pay more attention to strategic resources of firms and the way they are managed, more important being the ones having capability of providing economic value and competitiveness to the companies in the era of globalization and global outsourcing. Keeping in view this fact, it can be very well concluded that human resources and knowledge resources are the most valuable assets for organizations and an effective management of those assets boosts companies for an effective management of these assets to leverage knowledge and skill sets of employees. Conglomerates are building appropriate skill sets through various dynamic corporate strategies like acquisitions, partnerships, and tie-up, and cooperation in research and development. For example, Microsoft, Cisco, and Nokia have diversified their landscape through acquiring start-ups with Internet, software, human resource, knowledge management, outsourcing, technical and content capabilities.

Over the decades, the technological knowledge–based changes have led to changes in communication systems, transformation in knowledge management, digitalization of content, and social networking of knowledge, and now the world is moving toward global knowledge management system platforms like symbiotic web where humans and machines can interact in symbiosis.

# 2 Knowledge view of the firms

Knowledge is viewed as the most special strategic source that does not decline the way economic and other factors of production can; however, it increases exponentially through strategic planning and results in ultimately increased turnover of the firms. Most of the knowledge-based resources framework is intangible, dynamic, and lead to unique development through path dependency that forms the basis for economic rent creation and knowledge-based view of firms. The essential elements of knowledge-based view can be summarized as follows:

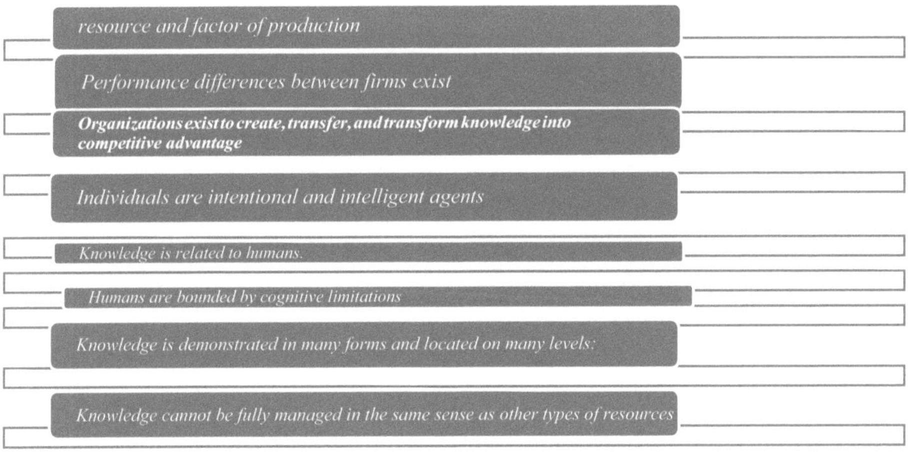


**Figure 1:** KBW of firms.

# 3 Knowledge management and web

In the era of globalization, the ability to manage knowledge is becoming increasingly more crucial. The creation and diffusion of knowledge are considered as important factors in competitiveness. Hence knowledge is being regarded as a valuable commodity that is embedded in products and in the tacit knowledge of highly mobile employees. Although knowledge is a valuable commodity or an intellectual asset, it possesses some paradoxical characteristics that are unique from the other valuable commodities. Knowledge cannot be treated as one thing, but rather as a diverse range of subjects which can be useful to someone in some situations. Companies are making their choices regarding the scope of programs and problematic within knowledge management. A three-part framework for categorizing the different types of KM initiatives cab be summarized as:

1. External structure initiatives (e.g., gain knowledge from customers, offer customers additional knowledge).
2. Internal structure initiatives (e.g., build a knowledge-sharing culture, create new revenues from existing knowledge, capture the individual's tacit knowledge, store it, spread it, and reuse it, and measure knowledge-creating processes and intangible assets produced).

3. Competence initiatives (e.g., create careers based on KM, create microenvironments for knowledge transfer).

The philosophy of KM 3.0 is focused not only on the individual; rather, it focuses on the use of knowledge. The advent of the Internet and the World Wide Web has made unlimited sources of knowledge available to us all. The growth of information on web has mounted with wider range of technologies, the information is sometimes not even recognized, considered to be complete and many times found inaccurate as well, making its access, acceptability, and usage not worth.

Therefore, to fit into this framework of knowledge management, the direction of which interlinks the identification, regeneration, processing, dissemination and applicability of knowledge to increase efficiency, growth and profitability in organizations seems to make advent of timeline of Web, Web 3.0. The major application areas of knowledge management are given in Table 1.

**Table 1:** Application areas of KM.

| |
|---|
| 1 Globalization of business |
| 2 Leaner organizations |
| 3 Corporate amnesia |
| 4 Technological advances |

# 4 Landscape of knowledge management

As there are different schools of thoughts for empiricism and rationalism pertaining to knowledge based on prior knowledge, experience, innate thoughts, perception, induction, the concept is more rational and useful in application for human behavior, desire, attitude, trust and interpretations. The thought process from various schools of thoughts has been used in knowledge framework, for analyzing it in a logically rigorous manner.

When we talk of constructing knowledge creation for an organization, a school, or a classroom, any platform of teaching or learning science needs to consider the nature of knowledge to be taught. Recent studies emphasize that scientific knowledge practices may not be differentiated in simplistic unitary way; that is, there is no single aspect of science, and that there are some hardcore commitment practices associated with scientific practices and knowledge without any implication for science. Epistemologists by and large least focus on procedural or acquaintance knowledge, however instead preferring to focus on *propositional* knowledge.

The school of thoughts may differ in the types of knowledge, so mutually propositional knowledge may claim to have different things with different variations.

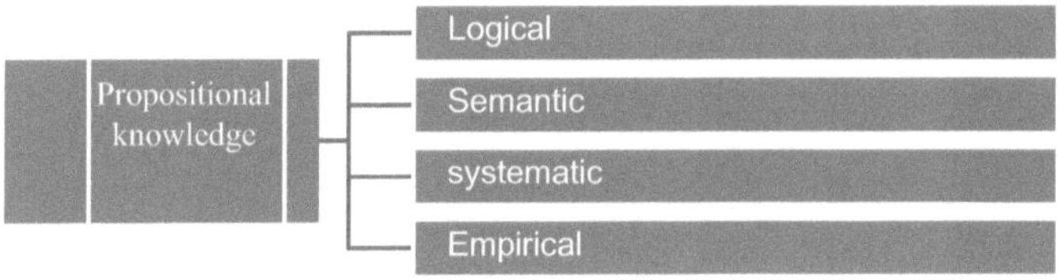


**Figure 2:** Propositional knowledge.

Epistemologists relate themselves with a number of tasks that may be sorted into two categories.

1. First, we must determine the nature of knowledge that someone must be able to judge the knowledge by a person, he succeeds in understanding or fails.
2. Secondly, we must be able to determine the extent of knowledge that how much, how can, how far we can know. The concept also emphasizes our testimony of others, our senses, our reasons, causes and resources to acquire knowledge.

Even if we restrict ourselves to active usages, there exist multiple senses of Knowledge and it may be imperative to differentiate between them. The difference in individual knowledge and collective knowledge can also be made. Social epistemology is the area of epistemology that addresses the way that groups, institutions, or other collective bodies might come to acquire knowledge.

***Knowledge and innovation*:** Innovation begins with construction of new kind of knowledge within the firm; this means that business re-organization/re-orientation is an essential precursor to innovation. Knowledge management has a significant role in identification and front-end market management. Organizations appoint chief knowledge officer to manage company's knowledge in the similar fashion in which CFO manages the capital. Without a formal system and a well-designed hierarchy, it's not possible to manage this process. Innovation and knowledge are nowadays two key activities for organizations. Innovation is one of most vital components in a knowledge-based economy. Both innovation and knowledge go simultaneously to create customer demand and new market areas.

# 5 Knowledge by acquittance vs description

While clearly distinguishing between knowledge by acquittance and description [20] mentions former having cognitive relation, being aware of the object ourselves while latter being constituting relation of judgement. The distinction between knowledge by acquaintance and knowledge by description is arguably a critical component of classical or traditional version of foundationalism in real senses. This is important as this may lead to illumination of the role that is played by acquaintance in securing

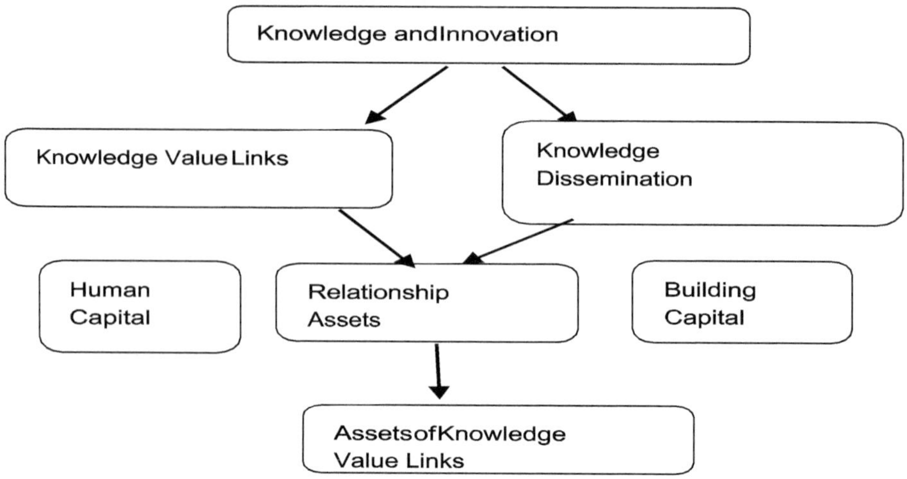


**Figure 3:** Knowledge and innovation.

knowledge, which in any case requires a justified belief and this may be the most supporting role when we fall short of knowledge [10].

**Table 2:** Acquaintance vs description.

| Knowledge by acquaintance | Knowledge by description |
|---|---|
| Relation of direct awareness | Ambiguous, indefinite |
| Without any process of inference | Involves process of inference, description |
| Non-judgmental, non-conceptual, | Descriptive only |
| Representational in Nature | Application of concepts |
| Creator of real genuine relation | Creator of judgmental descriptions |

Many experts have discussed that experience is the real provider of immediate "knowledge of acquaintance," "knowledge about" is the result of the systematic process that removes the independent and circumstantial contingencies of experience and extracts the principles that lie behind the "knowledge of acquaintance."

## 5.1 Explicit and tacit knowledge management

Explicit knowledge consists of facts, relationships, rules, and policies that can be easily codifiable in electronic form and shared without need of for discussion, while tacit knowledge, however, defies recording. This encompasses personal skills and its transfer requires face-to-face contact or even apprenticeship. The features are as displayed in Table 3.

**Table 3:** Comparison explicit vs tacit knowledge management.

| Features | Explicit | Tacit |
|---|---|---|
| Organizational results | High-quality, uniform solution to most of the problems | Appropriate, creative solutions, future opportunities exploitation |
| Risks and costs | Current risks and costs | Future costs and risks |
| Knowledge management | Codification, classification of explicit knowledge reuse | Identification of relevant and helpful staff |
| Targeted professionals | Teamwork, group formation innovators | High paid innovators, creative solution provider, one-to-one training for mentoring |
| Tools and techniques | Data analysis, policy documents, guidelines, risk assessment tools | Emails, meetings, workshops, video conferencing, road shows, coffee areas, case meetings |
| Source of knowledge | Professional analysts | Creative workers |
| Level of IT knowledge | Multiple investment reuse knowledge | Medium-to-low-level investment |
| Staff incentives | Staff adherence to polices recognitions, rewards | Emphasis on experts and original solutions |
| Organizations | Mainly service industry, customized good provider | Consultancies, product teams |

The link within the explicit and the tacit is evolutionary in that the selections made by individuals are settled in or out according to their utility in a particular historical and economic reality, and eventually embedded in organizational routines which then shape and constrain further individual choices.

## 5.2 Attitude toward inclusion

Knowledge vs Experience: The most successful inclusion programs are the ones where people are involved and influenced by the giver. Memory research of the past has tended to concentrate on memory performance, and to neglect memory as conscious experience. This neglect may reflect the acceptance of a tacit aspect that behavior, knowledge, and experience are closely correlated. The construction of scientific knowledge is a slow, continuous process of transformation taking place over a long period of time, involving sequential, uninterrupted estimations of becoming everyday knowledge. Leaning through knowledge takes mainly three scales or parameters for changes:

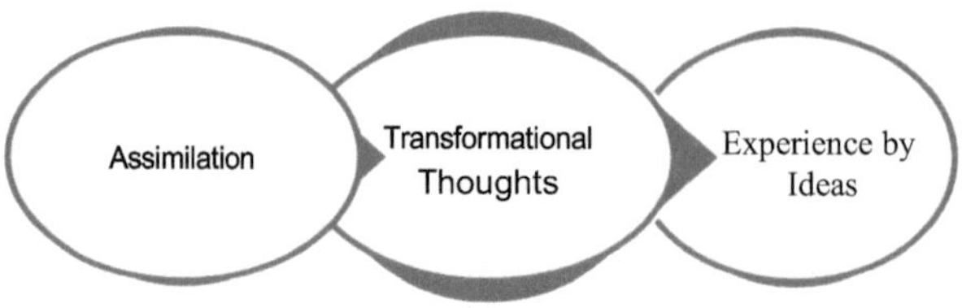


**Figure 4:** Knowledge parameters for changes.

- In most of the common cases, the gainers, knowledge learners, add on their already existing theories and practices.
- In other cases, any cognitive shock, experience, happening all put together lead the learner to deviate knowledge bank, attitude, and thought process.
- In the third case, although rare, assumptions are transformed by way of seeing, conceiving, and talking about from everyday knowledge.

Based on above, there are three implications for interactive designer.

The same can be summarized as,

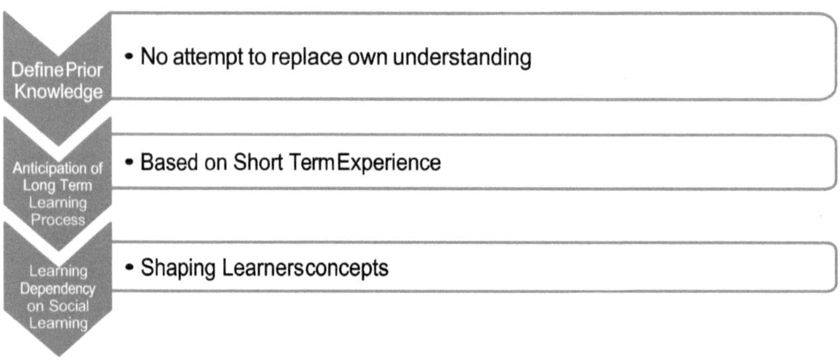


**Figure 5:** Interactive experiences implications.

It can be interpreted from above that only a part of knowledge that is specialized can exist explicitly as information; the rest is drawn from the community as a practice of disclosure. The importance of time, social parameter, and values change when it is being considered that the different methods experimented by scientists to gain knowledge. Properties of scientific knowledge arise from the social practices enacted by specific scientific communities. Discourse processes transform prior knowledge into refined concepts that can be applied consistently by members of the scientific community. Scientific knowledge is something like a polished commodity, for which prior knowledge supplies the raw materials and social interaction supplies the tools. Prior knowledge may be a multidimensional construct that when addressed independently does influence individuals' search for vacation information; and source credibility is the strongest predictor of type of information sources used [6]. The prior knowledge that an entrepreneur possesses plays a critical role in identification of entrepreneurial opportunities. Opportunity recognition is achieved through prior knowledge and prior experience.

Technology must understand relevant prior knowledge in a target user population to ease out its adoption. Prior knowledge is the most common factor for success and the most effective usage and technology has recently begun to explore the way challenges have to cope up aging population [13]. The boundary between the explicit and tacit types of knowledge is both porous and flexible, so there is traffic between the domains. The firm provides a special background wherein both explicit and implicit bodies of knowledge are both selected by mutual interaction to the external economic reality and then stored in the routines available to future generations of employees. Over time the quality of the interaction involved in explicit and evolving implicit types of knowledge may lead to further improvements and, thence, to superior firm performance.

## 5.3 Constructing scientific knowledge

There is another view that knowledge cannot be transmitted but must be constructed by the mental activities of learners on perspective of mental health. Not only personal but social perspective on learning as well as perspective on nature of scientific knowledge to be learnt is essential for presenting scientific learning in a formal setup. The organizational psychology is one of the biggest contributors of international cooperation and development of projects in multinational organizations and among professional community. Highly cited scientific production occurs within the communities only and the cohesive collaborations of similar knowledge domain and across communities. Organizational psychology may contribute to the implementation of in preview of social identity, manager adhere to professional identity, knowledge creation and accommodate professional norms and values which shapes up a knowledge led society. Ultimately management also efforts toward developing a knowledge-based identity where creativity and expertise hold important position.

# 6 Review of literature

Knowledge management diversified as a visible management topic by the mid-1990s as the topic gained momentum in the IT industry, application in computer services, and IT systems. During the last decade, although it has gained high popularity and visibility, with the advent of new technologies and new programs the area has come in for wide scrutiny as there are questions on the success of many of its applications, attempts, and especially those from the pure IT fields [8]. However in argument to this thought, many management experts have portrayed a new version of knowledge management as a field that evolves with the upcoming generation that both resolves the limitations of the last generation and brings more changes with new features that are beneficial to all.

There has been an exponential growth of online platforms, virtual areas of IT applications worldwide that are making it a tougher game for IT organizations to encourage a collaborative approach in sharing and leveraging their intellectual assets [1]. New-age technologies such as artificial management, IoT, and cloud computing have developed an interest area among organizations to develops such technologies of knowledge management alignment with the next generation with focus on age technologies. Knowledge management is still a young discipline and can be viewed in four emerging and complementary ways: ontological, epistemological, commodity, and community aspects [7] As knowledge management is an emerging, diversifying, and evolving stream that changes its applications, platforms, and features with emergence of every new generation, therefore, it's very natural to have even the contradictions in different aspects. Further adding to this are opinions that the challenges to knowledge management can be put in framework *engaging, exploring, experiencing, emerging, enabling and evolving* aspects of knowledge [12].

Knowledge management can be treated as a subject with more interlinks with literature, discussion, planning, and a part of action, and it is mandatory that for effectively implementing a strong and knowledge-based strategy, the organization must have strong knowledge pool; this is the primary success condition of knowledge implementation in an organization [2]. Despite such criticality, knowledge management remains an ill-defined term, as many other theories, technologies of management re-orient its application and part from this some technical theories here no categorization is being found are forcibly put under this stream. It is empirical to mention that the wider range of knowledge management has assisted organizations to have better insights into knowledge options, like applications and technologies in the related field. Various knowledge theories can be viewed for the totality and complexity, as it offers a wider range of tools and techniques. Knowledge theories can provide a framework within which an organization can balance its knowledge focus and establish its strategic knowledge direction with effective communication.

While relating leadership with knowledge,it is becoming mandatory and need of the hour for leaders of successful organizations to constantly search for better ways to improve performance and results which are knowledge-based [21]. In addition, past disappointments due to various management-driven initiatives have motivated managers to gain understanding into complex issues such as knowledge, which is a key governing area of any organizational effectiveness. Knowledge management is such a new aspect that it covers most of the cross-functional areas and has become a fashion in modern enterprises. The enterprises must create and sustain a balanced intellectual and human capital resource portfolio so that they remain competitive and successful on a long-term basis. The leadership in the organization must set broad and integrated goals for managing intellectual resources and it requires systematic knowledge management. With knowledge as the major leadership force behind the "generation of ideas," one can imagine that the emphasis must be:

- *knowledge creation*
- *knowledge development*
- *organizational knowledge*

*The above trio will continue to be the prime focus for improving society.*

There have been books, journals, and articles on knowledge management systems, but all the information available is in the form of surveys only [7]. Nevertheless, knowledge management systems (KMS) are the tools and techniques that support knowledge management practices in organizations, but only surveys and data interpretation cannot present this wide area as a cumulative knowledge bank. There must be some technologies, current landscape, content, tools that provide a framework for research into the development and use of these systems in organizations and the technologies only support the knowledge transformation hugely and future trends are also based on them only [5]. The explicit knowledge contributes strongly to technology solutions along with search and classification. The formation and communication of tacit knowledge, along with making it explicit, are some of the major some encouraging developments in knowledge transformation process. The attention toward the new term *technology-enhanced learning* (TEL) and related it to knowledge management (KM) emphasizing it to be an important means to increase individual and organizational performance and further highlighting the traditional KM and TEL models have failed to cope with the fast-paced change and critical challenges of the new knowledge era [14]. Over the past few years, the learning management system (LMS) has become an essential element of TEL. The emergence over the past couple of years of Web 2.0 technologies has provided new opportunities for alternative TEL solutions, often referred to as TEL 2.0 or E-Learning 2.0. Referring to Web 2.0 on social networking and collaboration between the synthesizer and the user while Web 3.0 as an intelligent web or semantic web with technologies like big data, linked data, cloud computing, 3D visualization, to make passive learner into active learner in the learning process [15], describes, Web 2.0 was transformation from web sites to social networking. In addition, it comes with features such as, computing platform, desktop applications replacement with web applications, attributes like network as platform, continuous updated software, and consuming and remixing data from various sources. It can be concluded that Web 2.0 is entirely a learner-centric activity; therefore, teaching and learning process got amalgamation. The predictions of Web 3.0 vary due to the variations in the technologies which will make up Web 3.0 in its formative years. On the contrary [18] has challenged Web 3.0, e-learning 3.0: the first major challenge is due to the nature of Web 3.0, its vastness, interoperability, lack of server-side checks, less privileges control, and increased privacy and security risk; the second major challenge is the wider application of Web 3.0, if current industry, education scenario will be able to utilize the benefits of Web 3.0 when the struggle with the previous generations is still going on.

To stay competitive, dynamic, and strategic, it is critical to stay updated with the technology trends as the time progression happens, the maturity of web happens as well in its own unique way [19]. Web 3.0, which is a unique web experience where machine will be able to understand and interpret data in a similar fashion as human being emerged from Web 2.0 which was interactive Web and which in turn emerged from Web 1.0 that has only static informative characteristics [19]. Experts have further revealed that Web 3.0 is a worldwide data warehouse where any format of data warehouse can be shared and interpreted by any device over any network, but the evolution of Web 3.0 will bring forth new opportunities and challenges. Autonomous integration of data and services which increase the preexisting capabilities of Web services, apart from creation of new functionalities, is the major opportunity, while unauthorized access and manipulation of data, autonomous initiation of action [18, 19], and the development of harmful scripts and languages would continue to be the major challenge in this technology. However, as a precaution, for anyone in the data information industry, data management expertise needs to be ready and acquire knowledge arising from Web 3.0 technologies. Further, as Web has been running into the era of Web 3.0 equipped with the development of semantic technology, therefore, the management of enterprise that suits Web 3.0 becomes a key addressal area for modern organizations. In this dynamic business environment, an emerging generation of users based on social networking experiences and empowered to create, publish, and redistribute content may find the structures of the Web 2.0 services a user-centered approach [4]. Innovative solutions would be required in this shift of web environment, innovative approaches to training, new emphasis on the role of resources as learners in changing business environment and rethinking the underlying architecture Web based technologies [4]. The continuous evolution of the Internet has thrown opportunities and challenges in web-based education and learning [17] and the traditional version of web, that is, Web 1.0, started as a read-only medium; the next version Web 2.0 established itself as read/write medium. Now the currently evolving version of web, namely, Web 3.0, is said to be a technologically advanced medium which allows the users to read/write/execute and also allows the machines to carry out some of the thinking so far expected only from the human beings. In a short time, Web 2.0 and now Web 3.0 have created new tools and technologies for facilitating web-based education and ramp; learning [3, 17].

As Web is the fast publication research medium all time, organizations consider Web-based technology as the most asset for income generation and cost control in the organizations [3].Web 3.0 creates the potential for implementation of new strategies and new forms of professional approaches and today, it is possible to assume the trends of Web 3.0 development [3]. New organizational strategies require both awareness of new opportunities in networking and acceptance and understanding of new organizational strategies of managers. The evolution of Web 1.0 to Web 3.0 has divided the education world into [9]:

- *digital natives*
- *digital immigrants*

*Digital immigrants* utilize traditional classroom pedagogy and communication methods in the classroom They went to school and graduated without *Google* or *Wikis* and have closely witnessed this epoch of technology revolution in education. In this sense, *digital immigrants* primarily include educators, educational administrators, parents, and lifelong learners who have been forced to adopt technology for education and learning.

It can be concluded from the above views of experts that the evolution of the Web 1.0 to Web 3.0 in the last three decades has had a tremendous impact on the growth and the future of education and e-Learning. Web 3.0 or the "Semantic Web" builds on Web 2.0 or the "Social Web" which is characterized by learner interaction and participation on the web. As the third generation of the web, Web 3.0 uses data from learner interaction on the web to generate and deliver customized content to the learners. Computer algorithms form the basis of Web 3.0 which makes it possible for machines, tools, and application to exhibit artificial intelligence (AI).

# 7 Web transformation

The Internet was invented by Tim Berners-Lee in 1989, with the vision of communicating or sharing information in a common information space, called web. By the exponential growth, it became a medium to store and access various versions of learning contents such as text, pictures, audio, and video. The main challenge of Web 1.0 is the creation of the content and sharing with others. The users are not able to curate knowledge. As a result, the Web 1.0 is widely recognized as read-only web with its general attributes like static information and minimal interaction between web sites. As per sources, Web 2.0 is a transition from web sites. Added to that, it is a full computing platform, replacing desktop applications with web applications, with features like network as platform, continuous upgraded software, and consuming and remixing data from various sources. Digital Natives were able to curate content and it demands the learner to participate actively in the learning process. Web 3.0 predicts that online content is organized in a semantic way, much more personalized for each user, intelligent websites and applications, and search and behavioral advertising. Integration of Web 3.0 technology into KM 3.0 system to collect, share, and use valuable knowledge to help people get their job done. Therefore, KM 3.0 is about having knowledge we can use rather than collecting and sharing as much as possible [11].

The characteristics of all forms depicted in Table 4, as

**Table 4:** Characteristics of Web 1.0, Web 2.0, and Web 3.0.

| Web 1.0 | Web 2.0 | Web 3.0 |
|---|---|---|
| Read | Read and write | Read, write, and execute |
| Client server | Peer to peer | Portable personal web |
| HTML portals | XML, RSS | RDF |
| Companies own | Communities share | Individuals |
| Web pages | Web applications | Semantic applications |

# 8 Knowledge transformation

Many organizations accumulate and created large knowledge systems, expert systems, knowledge repositories, and intranets to ensure the availability of their knowledge assets. Without having a clear strategy for using this knowledge the companies at least made sure that they collected it and kept it in a database or file somewhere. KM 2.0 focused on sharing knowledge using web-enabled and social media tools: With the origin of Web 2.0, focus shifted toward sharing, communication, and collaboration of knowledge. The popularity of social media and web tools has started to reach the enterprise and the use of these tools in an enterprise setting is often named “Enterprise 2.0” [11]. The focus of KM 3.0 lies on using existing knowledge to help people gets their job done: this simply means having access to useful knowledge whenever required. In the research by [11] knowledge worker has been equated to an island – where there is less common knowledge and more specialized knowledge. It shows that the worker must collect more for themselves, and less for others. Each knowledge worker is also supposed to be responsible for reducing the noise and the information overload by focusing more on sharing quality than quantity [11].

Table 5 discusses the differences between KM 1.0, KM 2.0, and KM 3.0 [11, 16]

**Table 5:** The difference between KM 1.0, KM 2.0, and KM 3.0.

| KM 1.0 | KM 2.0 | KM 3.0 |
|---|---|---|
| Techno-centric<br>Command and control | People-centric<br>Social | Productivity-centric<br>Practical |
| Centralized systems | Decentralized ecosystems | Personal and decentralized systems |
| Email, newsletters, databases | Social tools (blogs, wikis, IMs) | Practical, personal, social, |
| KM is extra work | KM is part of my work | KM is helping me do my work |

*Source:* http://www.tlainc.com/articl256.htm [11, 16]

# 9 Discussions

We live in an era of information society which is based on information opportunities, where the object is not simply knowledge but the possibility of accessing it, producing, and diffusing new knowledge through global information media which a user can share, read, and writes through computers connected with internet web pages. Partial or exclusive focus on people, technologies, techniques, and systems may not enable firms a long-term sustainability but interaction between technology, techniques, and people can enable an organization to manage knowledge effectively. "Learning by doing" and "application-based management" kind of environment sustain organizational competitiveness. With these objectives, during the last few years knowledge management has extended to the adoption of technology, Web-based advancements, multiple-sensor systems, semantic web where computer is generating and thinking new information rather than humans. However, the transformation process of Web 1.0 to Web 3.0 is also accompanied with some of challenges like lack of data standard for controlling metadata and data privacy, resource description framework (RDF), web ontology language (OWL) used unified resource identifiers (URI) to represent data which can be held in database without specifying any access policy or trust boundaries. There can be chances that these challenges may make Web 3.0 vulnerable and attackers may also falsify the data internationally, and wrong coding, false services, and unwanted interpretation can happen. Data privacy is one of the most important key issues in Web 3.0. Although social and semantic Web have grown separately, both have a combined approach to leverage crisis management, formal reasoning capabilities, and decision-making in emergency situations. Nevertheless, it is unquestioned that the enormous pool of user-generated information contains pieces of highly relevant, applications, techniques, solutions, and information that are not known to the decision maker to the full extent.

Sharing of feeling using Web 3.0 technologies, creating, processing thoughts and ideas instead of linking documents only. Social web is considered as an efficient and attractive way of connecting people around the globe. The platform also allows people to find the information at a much deeper and extensive level with the meaning of the search terms and the context in which they are used. Web 3.0 enables the information in a way that machines can read it and understand it as much as humans can without ambiguousness. Today's generation live in a virtual world, participating in group activities. Web 3.0 provides applications that are fragmented together, are in cloud, and can be run on any device very fast, very customizable, and distributed virally. It has been witnessed that manifest in Web 2.0 and Web 3.0 has been a great extension of Knowledge Management, a true communal medium the distinction between professional, semi-professional and consumers has got aligned creating a network effect of business and applications.

The future generation of web is gearing for the next great phase of Web, where not only people (Web 1.0 and Web 2.0) or machines (Web 3.0) are connected but objects like objects toaster, to a car, our keys, our phone, books, geysers, and lights fans, the list is endless, the term coined as Web 4.0. It offers a new model of user interaction with the most comprehensive and personalized, not limited simply to display information, but proposes to behave like an intelligent mirror that concrete solutions to what the user needs.

The Web is continuously evolving. The access to the Internet is not made exclusively by humans, but also from other objects, devices, and vehicles, originating the concept of Internet of Things. These new concepts and demands have led to the birth of a new Web model, entitled Web 4.0.

The concept of Web 4.0 is quite complex and changeable. Till now, there is not a mutual consensus regarding a concrete definition of Web 4.0, contrary to previous wide-accepted definitions of Web 1.0, Web 2.0, and Web 3.0. However it can be said that Web 4.0 will now craft the landscape for future generations with its unique applications such as social networks and technologies such as Internet of Things, Big Data, AI, and M2M that will play a key role in the adoption and implementation of Web 4.0, the new Web that fosters experts for further research in the area.

## 10 Study implications

It is seen that to gain competitiveness and to improve business processes, number of organizations have included KM as a business strategy. It is also noticed that reports of successful KM implementations have resulted in financial savings, increased revenues, and increased level of user acceptance. The researchers found that sharing tacit knowledge is key for mentoring success, and express concern that remote learning, which is becoming more common during the COVID-19 pandemic, may not be an effective means of transferring knowledge such as tacit knowledge. Also, models help us to put the disparate pieces of a puzzle together in a way that leads to a deeper understanding of both the pieces and the ensemble they make up. KM models are still new to the practice or business of knowledge management, and yet they represent the way ahead. It requires a solid foundation of understanding of what KM is and what the key KM cycle processes are, and how these fit into a model that enables us to interpret, to establish cause and effect, and to successfully implement knowledge management solutions. The Web begins to bring more knowledge and action capacity for its users, resulting in considerable changes in several aspects of daily life. Web 3.0 is moving fast toward a more dynamic and faster-changing environment, where the democratization of the capacity of action and knowledge can speed up business in almost all areas. So, there should be some solution to transfer tacit knowledge using Web 3.0 or Web 4.0. Now Web 4.0 will be the Internet of things or we may name it as a web of intelligence connections digitally.

## References


[1] Abar, S., Abe, T. & Kinoshita, T. (2004). A next generation knowledge management system architecture. 191–195, 2. doi: 10.1109/AINA.2004.1283784. Abar S, Abe T, Kinoshita T. A next generation knowledge management system architecture. In18th International Conference on Advanced Information Networking and Applications, 2004. AINA 2004. 2004 Mar 29 (Vol. 2, pp. 191-195). IEEE.

[2] Binney, D. (2001, Mar). The knowledge management spectrum–understanding the KM landscape. Journal of knowledge management, 1.

[3] Bruwer, R. (Hendrik J., & Rudman, R. (2015). Web 3.0: Governance, risks and safeguards. Journal of Applied Business Research (JABR), 31(3), 1037–1056. doi: https://doi.org/10.19030/jabr.v31i3.9241.

[4] Craig, E.M. (2007). Changing paradigms: managed learning environments and Web 2.0. Campus-Wide Information Systems, 24(3), 152–161. doi: https://doi.org/10.1108/10650740710762185.

[5] Marwick, D. (2001). Knowledge management technology. IBM Systems Journal, 40(4), 814–830. doi: 10.1147/sj.404.0814.

[6] Kerstetter, D. & Cho, M.-H. (2004). Prior knowledge, credibility and information search. Annals of Tourism Research, 31(4), 961–985. ISSN 0160-7383, doi: https://doi.org/10.1016/j.annals.2004.04.002.

[7] Gallupe, B. (2001, Mar.). Knowledge management systems: surveying the landscape. International Journal of Management Reviews, 31, 61–77.

[8] Grant, K.A. & Grant, C.T. (2008, Jan, 1). Developing a model of next generation knowledge management. Issues in Informing Science & Information Technology, 5.

[9] Guo, R.X., Dobson, T. & Petrina, S. (2008, Apr). Digital natives, digital immigrants: An analysis of age and ICT competency in teacher education. Journal of educational computing research, 38(3), 235–254.

[10] Hasan, A. (2011). Classical foundationalism and Bergmann's dilemma for internalism. Journal of Philosophical Research, 36, 391–410.

[11] Iversen, A. (2009), KM 3.0 Part IV: a practical KM system, (Accessed on December 6,2020 at http://www.ppcsoft.com/blog/km-3-4.asp).
[12] Jakubik, M. (2011). "Becoming to know. Shifting the knowledge creation paradigm". Journal of Knowledge Management, 15(3), 374–402. doi: https://doi.org/10.1108/13673271111137394.
[13] Lee, C. & Coughlin, J.F. (2015, Sep). PERSPECTIVE: Older adults' adoption of technology: an integrated approach to identifying determinants and barriers. Journal of Product Innovation Management, 32(5), 747–759.
[14] Chatti, M.A., Schroeder, U. & Jarke, M. (April-June 2012). LaaN: convergence of knowledge management and technology-enhanced learning. IEEE Transactions on Learning Technologies, 5(2), 177–189. doi: 10.1109/TLT.2011.33.
[15] Blog, M. & Meichel, F., Qu'est-ce que la cognition – points de repères en sciences cognitives Qu'est- ce que la cognition – points de repères en sciences cognitive. Retrieved March 20, 2012 from http://florencemeichel.blogspot.com/2009/05/quest-ceque-la-cogntion-points-de.html.
[16] Mixotricha, Z. (2008), KM 1.0 KM 2.0 KM 3.0, http://zyxo.wordpress.com/2008/12/28/km-10-km-20-km-30/
[17] Rajiv, & Lal, M. (2011). BVICAM's International Journal of Information Technology, 3(2), 16–22.
[18] Ron, C. (2009). Web 4.0, Trip Down the Rabbit Hole or Brave New World? (Accessed on Dec 05,2020 at http://www.zmogo.com/web/web-40trip-down-therabbit-hole-or-brave-new-world/.)
[19] Rudman, R. & Bruwer, R. (2016). Defining Web 3.0: opportunities and challenges. The Electronic Library, 34(1), 132–154. doi: https://doi.org/10.1108/EL-08-2014-0140.
[20] Russell, B. Knowledge by acquaintance and knowledge by description. In Proceedings of the Aristotelian Society 1910 Jan 1 (Vol.11, pp. 108–128). Aristotelian Society, Wiley.
[21] Wiig, K.M. (1997). Journal of Knowledge Management, 1(1), 6–14.

Pradeep Bedi, S. B. Goyal, Jugnesh Kumar, Ritika

# Integration of Knowledge Management and Web 3.0 for Human Resource Management

**Abstract:** Knowledge management (KM) is a technique that is used to develop and manage organizational performance in such a competitive scenario due to current fundamental economic resources. This chapter is aimed at analyzing KM linked to decision support systems for human resource management to argue the relevance of knowledge in organizations. The implementation of KM has increased rapidly with the development of Web 3.0. So, the management information systems (MIS) of an organization must be upgraded and information stored in it can be made available easily to those who need it. The proposed framework will deploy KM to achieve high performance and to achieve a sustainable advantage. This framework will focus on the organization's functional structure, operational and marketing aspects, and technical aspects.

**Keywords:** Knowledge Management, Knowledge Resource, Decision-Making System

## 1 Introduction

In an organization, human resources are considered to be one of the most prominent parts that are needed to be monitored for growth and development. Therefore it is considered to be a typical management task. Management of human resources is a hectic task that includes functions such as hiring staff for the organization, tracking employees' performance, developing employees' skills, compensation of employee efforts, and providing requirements of employees. In human resource management (HRM) adequate tools and techniques are also required same as in another domain, to resolve the tasks that can reduce the human efforts and time consumption with accuracy. HRM uses a broad set of heterogeneous tools and techniques to improvise and professionalize entire work. Apart from HRM tools and techniques, other tools from other domain are also adopted as they are required for the analytical process of information such as psychometric analytical tools which are used to analyze the sentimental and emotional state of employees. So, apart from data analytics, this is one of the important working parts of HRM that can be successfully achieved from information systems using Web 3.0 [1–5].

Another important uprising tool is artificial intelligence (AI) or machine learning that can be applied in HRM. In this chapter different AI approaches for knowledge management (KM) using AI or machine learning are discussed. There are a

https://doi.org/10.1515/9783110722789-005

large set of heterogeneous AI techniques to handle specific HRM tasks such as product improvement in industries, smart employee selection process, smart employee recruitment process, and performance evaluation of employees. These contributions have given a broad insight into the potential of AI techniques to explore HRM tasks and to improvise their efficiency. Thus, this chapter explores the concept of Web 3.0 and its applications in KM as well as technologies such as AI and cloud computing and their proficiency in HR management. In Section 2 overview of KM is discussed with their application areas. Section 3 discusses the enhanced information management (IM) for understanding issues of IM and their enhancements. Section 4 describes an overview of Web 3.0 with their needs and objectives with their tools. Section 5 describes the application of Web 3.0 in HRM in the current scenario. As in Section 6 KM and its integration are discussed. In Section 7, the application of AI for KM is discussed. In Section 8, after analyzing the drawbacks of the existing system, a framework for HRM integrated with the machine learning approach and cloud computing is proposed. Finally, in Section 9 conclusion and future research scope of this chapter are described.

## 2 Overview of knowledge management

To have success in this real world two things are very crucial for the company: the first is innovation and the second one is performance; to stay for the long run in the global market. For that, it is very important to stay more strongly than ever before to meet the customer's satisfaction, thereby resulting in the increase of productivity and competition. Following the changes in the field of technology must also be needed to maintain an adequate contribution of every tool as well as techniques along with the accessibility of information [6]. This knowledge was diversified, delocalized, and not easily managed, leading to the development of several methods for controlling information systems to leverage it as soon as possible. This information is currently processed and handled by taking its context and its semantics into account, which means that we handle awareness. Nowadays, information is seen as a strategic resource and a stability element, giving it a decisive competitive advantage. To innovate in products, procedures, services, and the organization, KM is essential for the business. It allows its design costs, development, delivery, and so on to be minimized at the same time. Managing understanding is not a new problem. The difference from the past is that businesses are now explicitly attacking this topic according to an aware, regulated, and voluntary approach [7]. In the past, this was done without thinking about it. KM appeared as a discipline in response to a wide range of problems arising from memory losses due to retirement departures, technological growth, and innovation. To improve the added value of design and development methods, it is seen as a practice that needs particular approaches. To improvise the skills and to increase the profits of an organization, different companies adopt different strategies. All strategies are designed to maintain a proper place or rank in the market. Approaches must be different, but these all ultimately lead to the profit earning for the organization. However, for designing these strategies, KM is quite necessary that leads to the exponential increment of profit concerning investment cost. So, KM can be explored differently and dynamically. KM will lead to handling long-term challenges such as capitalization, team co-ordination, recruitments, retirements, performance evaluation, and project management. This can be achieved from past learning and their deployment in present conditions to achieve the dedicated mission and vision of an organization. In current competitive world, IM guarantees the company's survival [8].

As a consequence of the rapid changes in the business environment today, the need for IM has increased. Second, customization systems require knowledge of different consumer needs and expectations. Second, KM, which involves codification, personalization, and information process controls, requires numerous end-use technology implementations and acceleration of technological change. Third, the increasing variety of information sources from the increased use of value-chain outsourcing and deconstruction demands that these increased sources be controlled. In recent years, IM has gained widespread attention. In the current competitive environment, the value of information is of the highest priority in every field, be it the industrial or the educational field. There is a huge contribution of researchers to explore the advantages of KM for understanding the information value of an organization. So, to further explore the advantages of KM it is quite required to first understand the need, objectives, and tools of KM in the exploration of knowledge of organizations. In this section, definition and principles of knowledge management (KM) is summarized [9].

A multitude of interpretations and meanings of information and KM exist. It is important to identify them before proceeding further:

- Knowledge is a whole collection of customer-related instincts, logic, observations, interactions, goods, procedures, economies, competitiveness, and so on that make for successful action.
- Knowledge management is a structured, coordinated, clear, and intentional mechanism in which knowledge is developed, disseminated, implemented, refreshed, and modified to achieve organizational goals.

Two main facets are required to understand the definition of KM is such as:

- Treating the information aspect of company operations as an explicit market issue reflected at all levels of the enterprise in strategy, practice, and policy.
- Establishing a clear correlation between the intellectual capabilities of an entity, both overt and implicit, and development. In reality, KM also includes defining and mapping intellectual properties within the enterprise in terms of these two factors, creating new expertise within the organization for strategic advantage,

making large quantities of organizational information available, exchanging best practices, and technology that makes all of the above.

Unless functions of KM are specified, its utilities cannot be observed. To understand the functioning of KM it is required to understand their values in order to withstand in this competitive scenario. These values are discussed as below:
- The most important value of KM is awareness that needed to be connected. In perceptions and insights, it is the collective knowledge and its value stems from its relational relationships and qualities surrounding its content.
- Compatibility of KM with this dynamically changing world is also another important value that needed to be achieved. This can be resolved by integration awareness about new information, competition in market as well as development.
- It is termed as catalyst which improves the activities or actions of an organization. For this value awareness is also important as it tracks the environmental conditions and tracks the dynamic development of the market.
- The solution of KM will also contribute for information sharing. The variation in this concept will inform the organization about the changing activities of market and competition.

Along with these values, different analysts have set up their theories to establish relationship between KM and IM [10].

# 3 Enhanced information management

To study the enhancement of KM, first of all, it is needed to understand the relationship between IM and KM. As in an organization, several pieces of information, in the form of emails, manuals, surveys, and so on, are being performed to improve and make the expertise more analytical. As it is a quite difficult and hectic task to share knowledge individually due to various reasons such as restriction of time and geographical location a collective automated platform is required for all individuals working in an organization. Therefore, to achieve expertise analysis of KM, it is quite necessary to integrate with IM.

This information directs the employee to the appropriate source of expertise; it doesn't matter if it is a text or an excerpt. These maps of organizational information or directories of expertise identify a collection of categories of knowledge, the position of the knowledge, and, in some cases, its state and value. It exists as an entity to fulfill various unique tasks and goals. With improvement or enhancement of information systems, information use and reuse can be successfully performed by expertise, which is encouraged by KM. This complex distinction between KM and IM focuses on the related component of information as a vital distinguishing attribute [14–18].

# 4 Overview of Web 3.0

While understanding the features of Web 3.0, it is necessary to understand the transition of Web 1.0 to Web 2.0 and from Web 2.0 to Web 3.0. It is because all features of Web 1.0 are integrated into Web 2.0 as well as all features of Web 2.0 are contributed into Web 3.0 (Figure 1). In terms of information, in the Web 2.0 era, several networking networks like Blogger, Twitter, or Facebook have been created and the sharing of data between multiple users has shifted. The new Website update, known as Web 3.0, incorporates the characteristics of both phases with some more features (Figure 1). By Web 3.0 we mean an alleged third wave of Internet implementations, which together constitute what could be referred to as "smart web" including semantic web users, microformats, natural-language analysis, data storage, computer learning, AI technology. "Web 3.0 is the third generation of Internet-based web services that emphasize machine-facilitated knowledge of the information to provide a more productive and intuitive user experience," says Wikipedia. Web 3.0 is one of the leading technology for digital era as stated by Nova Spivack, CEO of Radar Networks. (2010–2020) [19, 20].

In other words, the third wave of the Internet that is made accessible by a mixture of different new technologies can be explained as follows [21]:

- Transforming the site from a transparent and interoperable network with diverse apps and content repositories.
- Universal networking, cable network penetration, and smartphone Internet usage.
- Network computing, business models for SaaS, online service interoperability, remote computing, cloud computing, and fog computing.
- Open technology and frameworks along with APIs and protocols.
- Open identification, open credibility, smartphone identity roaming, and personal details.
- Smart mobile, web semantic technology.
- The "World Wide Database" distributed database.
- Smart software, natural language, artificial learning, machine thinking, and automated agents.

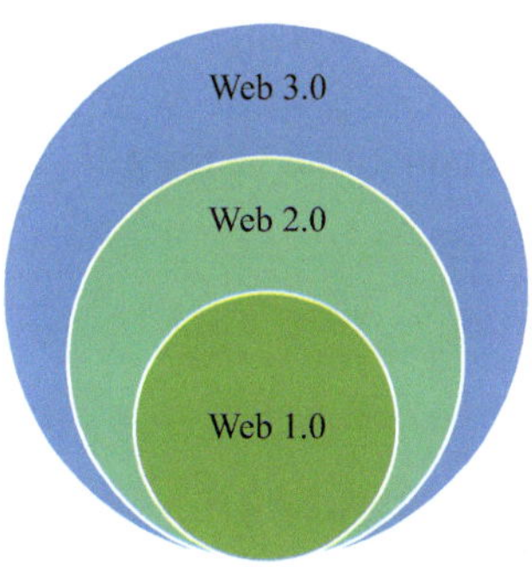


**Figure 1:** Evolution of web.

## 4.1 Need for Web 3.0

In Web 1.0 or 2.0 when anything is searched over the Internet then it will display some website links that may or may not be acceptable by an individual. Then in such a condition he/she has to enter different keywords to access the appropriate information. So, such web searching is dependent on keywords. Such implementation takes a quite long time to get the appropriate result. So, there is a requirement for improvement in the searching capability of web applications that can take less time to give a better solution. Web 3.0 overcomes this challenge when it lets the user get the answer he needs. Network 3.0 is a way to fix the issues described above. The whole quest mechanism can be shifted as platform 3.0 is integrated into the system. This will help to gather information easily from the Internet. By searching, arranging, and only presenting the pertinent information the system can be better equipped to provide the user with the information needed. An "age of artificial intelligence" can be called the Web 3.0 era.

## 4.2 Fundamentals of Web 3.0

Certain analysts say that Web 3.0 has almost every piece of information about a remote wizard. This is why it is often referred to as a "giant archive." Web 2.0 uses the Internet to link people, while Web 3.0 is used for knowledge sharing Web 2.0. The following example will help illustrate this. You need to look at these places using the latest apps online to get the information you need on travel and sites, visitor destinations, and so on if you want to get a holiday with an approximate budget of $3,000. The prices of various travel agencies can be compared by visiting numerous websites. This is undoubtedly a time-consuming and boring approach. But the time spent browsing multiple sites can be reduced with Web 3.0, so the Internet does all this, through narrowly defined search parameters. The Web browser can compile, interpret, and display the data in a manner that makes it easy to compare. This happens as the information available on the web is understandable by Web 3.0. A Web 3.0 search engine scans not only to search for the given keywords but also interprets their meaning. It can include related findings along with recommendations for more data searching. The "Internet of services" is often referred to as Network 3.0. In addition to the human-readable web portion, open machine resources can be merged into higher service standards.

## 4.3 Structure of Web 3.0

In contrast to the previous ITs of the Internet, Platform 3.0 has a new GUI. Semiotics and semantics describe the fundamental framework. Semantics refers to communication meaning, while semi-organisms are the study of symbol processes and signals (semiotics), both independently and organized into sign or context and communication structures. It needs to analyze how meaning is formed and understood. Semantics can benefit the consumer very well, as it allows him to get the right answer.

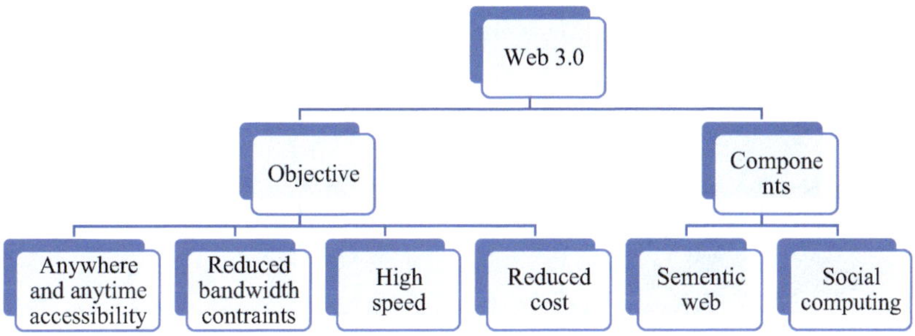


**Figure 2:** Objectives and components of Web 3.0.

## 4.4 Advantages of Web 3.0

Improves data management: Data administration on the Internet is a complex activity since it requires the interpretation of many programs for the convergence of the data of various systems that can or may not be interpreted by a computer in certain cases. In these cases, the relation of data to the appropriate input is very difficult. This dilemma can be solved to a degree by using the semantic web if it tells the mutual understanding between the various data sets, making it easier for the system to understand and efficiently combine connections with the desired output.

Mobile accessibility: Overall, by the end of 2008, more than 4 billion mobile device users was anticipated. The worldwide smartphone penetration rates were expected to exceed 50% at the same time. But with the launch of the 3G technology, this number was projected to hit 60%, making it easier to access the Internet on handheld devices. As a result, the market for Web 3.0 is projected to grow due to its excellent performance. Internet access via mobile devices is supposed to have an important part to play. The size of the web page [22–24] in Web 3.0 can be reduced by the application of CSS.

Creativity tools: The multiple links between separate databases is the main feature on Platform 3.0. Structures and individuals will use the data and information datasets in the same manner and thereby boost their effectiveness. This facilitates innovative processes, which in turn lead to the development of a new model for ideas and R&D.

Globalization: Network 3.0, with a common programming language called the resource definition framework (RDF), is attempting to construct and standardize various data structures. This is the underlying data exchange model of the network.

RDF is written in XML such that different programs can easily exchange details. This breaks constraints on latency, poor web view on mobile devices, and so on.

Enhanced customers' satisfaction: Network 3.0 gives the corporation "artificial intelligence" to increase the degree of customer service satisfaction (CSS). Via this, consumers can view the full product information on their webpage. This increases the reputation of the company on the market, leading to more customers. Some of the technologies employed in Web 3.0 are illustrated in Figure 3.

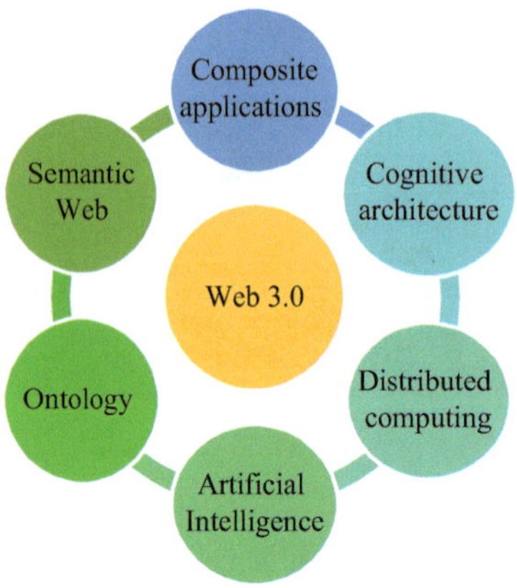


**Figure 3:** Technologies of Web 3.0.

# 5 Web 3.0 and HRM

Each field is being changed by launching a better plan that improves with the matter of time. The convention of a particular system has been enhanced by Internet of Things (IoT). An example of frond end applications of IoT is smart gadgets that are adaptable sensor by using its features. Periodically, the gadgets of IoT with the capacitance of heterogeneous billions there is implementation and constraints quality need of quality of service (QoS). HRM has grown to the advance to all of things movement, business capacity, heterogeneous data support, drawing practicality, data integration, and high relativity in the year 2020 [25].

The tools and services provided by the network infrastructure and the Internet will be accessed by human resources professionals. In the furtherance of Internet software, the concepts Web 1.0, Web 2.0, and Web 3.0 have typically been accomplished in many stages. Web 3.0 was planned as the development of high-quality goods and services by experts using Web 2.0 technology as a feasible platform, by Jason Calacanis in 2007. The true implementation of technologies and software devices started in 2010. There is no competition in a useful interpretation of instances of the web 3.0. Therefore, many mobile or smart applications used the concept of web 3.0 to deliver better video quality, machine control tools and many more 3-D effects including augmented reality [26].

In current internet scenario it is quite difficult to distinguish between Web 2.0 and Web 3.0. The introduction of Web 3.0 or Web 2.0 in human resource management systems (HRMS) is in wide application for human capital management, workforce management, talent development system, business learning systems, e-recruitment, etc.

With application of Web 2.0 or Web 3.0 technologies for applications in HRM, this would motivate the participation of users to interact with information and enhance services for HR professionals. To support this, there are already many HRMS applications are available [27].

Currently, many smart applications are also available as a separate online HR support. In the HR management, there are scope for research opportunities for developing tools and techniques with integration of Web 3.0.

# 6 Knowledge management and integration (KMI)

In the last two decades, KM has gained a lot of attention to sustain in this competitive world. This has increased the ability of an organization to focus and embed specialist knowledge to improvise their outcomes. Knowledge Management and Innovative technologies had shown up the proficiency for the organization in terms of profit and satisfaction level [28]. After studying the previous contributions of researchers in the field of HRM, KM is considered to be the most important pillar for information extraction and their management processes such as knowledge acquisition, sharing, transfer, and integration. A general social scenario is created by the HRM practices which move in the direction of proficient knowledge integration. Before understanding the relation between HRM and KI, it is important to know about goals (Figure 4).

In recent studies [29–33] relationship between HRM and KM processes have been explored along with performance results. So, these studies have improvised the understanding of the role of KM in every field. It is also known how the HRM is influenced by the KM for other firms such as outsourcing firms. This shows the raised importance of KM from the last two decades which helps in creating a knowledge base for the deployment of new products and processes that contribute to increment of productivity as well as profit (Figure 5).

# 7 Contextualization of knowledge management using AI

Suriyan Jomthanachai et al. [34] showed the HRM application in rubberwood factory for converting it from industry to industry 4.0. The main objective of conversion was the large number of production data that are quite difficult to handle manually.

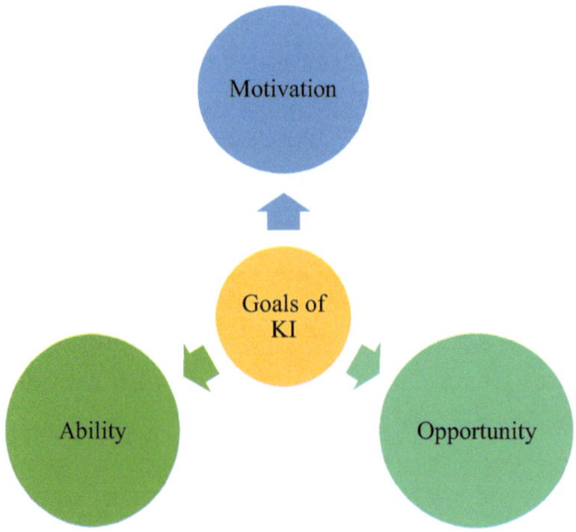


**Figure 4:** Goals of KI.

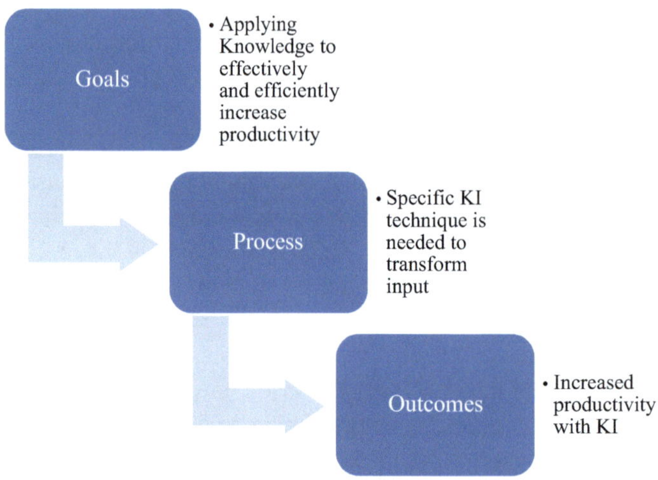


**Figure 5:** System model of KI.

This causes incorrect compilation of data and delayed services. This serious issue faced by the company also affected the inventory management as well as salary payment of employees. This ultimately leads to economic issues for the company along with the labor-intensive tasks. Then to resolve these issues, the application of machine learning (genetic algorithm (GA)) was used. The application of GA checked utilization of raw materials and also performed the data analysis task with efficiency. A web-based application was also developed to manage such large production data and analysis to enhance the productivity of the entire system timely.

Anna Sołtysik-Piorunkiewicz et al. [35] studied the usefulness of web technologies and their support systems to manage data in organizations. The application of KM was also studied in this work. In this study, a detailed discussion for stages of KM was discussed such as generation, collection, analysis, and evaluation of information or knowledge for application with new-era web technologies. The technique was focused on the structural, functional, technical, and analytical aspects.

Maria Dominic et al. [36] discussed technologies and features of Web 3.0 such as semantic web, big data, cloud computing, and augmented and virtual realities. The study states that the development of Web 3.0 has moved from passive learning to active learning technologies. In this work, the e-learning area was focused on observing the impact of Web 3.0 features and technologies and along with that different issues associated with Web 3.0 were also studied.

Prof. Mohamed M. El Hadi et al. [37] discussed the features of Web 3.0 such as semantic web, big data, cloud computing, and AI. Author discussed the contribution of AI and semantic web for successful and efficient knowledge management. In this work, educational KM is taken as the domain to show the proficiency of Web 3.0 with semantic web and AI features. This makes the process of learning quite varying from traditional processes normally utilized in teaching and learning. This work presents KM systems prototype in educational organizations by utilizing Web 3.0 technologies.

Amit Sheth et al. [38] discussed methods to utilize all types of modern web resources utilizing the key principles of the capability to relate meaning with data via conceptual or domain models and semantic details consisting of annotations, and through advanced semantic techniques for search, integration, and analysis. This work talks about the utilization of semantic web standards and methods but also supports the utilization of lighter-weight, convenient, and more scalable choices when they are more convenient.

Robin D. Morris et al. [39] discussed about the impact of Web 3.0, semantic web, on the e-learning platform for its development. It is said that the online users must have the knowledge of Web 3.0 technologies for upcoming advancements in the field of e-Learning. This work gives an illustrative understanding of semantic web and its features as the basics of semantic web are unification of data. In Web 3.0, by utilizing the metadata data is transformed into significant details that can be accessed and analyzed. This work helps the online users for e-learning such as tutors, for evaluation of course and in learning.

Víctor Hugo Medina García et al. [40] discussed the association and interconnection between complicated theory and KM. Planning was done to find processes to exchange or share knowledge effectively, upgrading KM in a complicated organization like the evaluation of software factories. As a consequence, it gives a prototype of KM, considering the concept of complexity theory employed to a formulation of system software factories, where usual formulation requires various types of complications supervised properly.

C. Rosa Yeh et al. [41] discussed that how technology had changed the process of acquiring skills. This work was carried out to assist entrepreneurs to better understand the varying technologies and its impact on society. From the point of connectivism, this work is intended to integrate the advanced learning approach. Connectivism study includes the social interaction technologies and implementation of KM. Empirical overview questionnaires were redirected to 301 industries in the computer firm covering Taiwan, out of which 80 authentic feedbacks were obtained. The hypothesis was proposed that proved the direct impacts of corporate culture on organizational learning. Newly established industries show comparatively elevated enthusiasm for innovation, learning, and its adoption of modern technologies.

Tsang-Kai Chang et al. [42] demonstrated the hypothesis to project and understand actions as well as behaviors over the Internet. The e-learning domain was taken as an example. In this work, experimental analysis was performed to extract the impact of users or contestants to improve the usage caliber over the e-learning platform. The experiment was performed on 176 educational administrators. A small opinion polling was performed to understand the "superior effect," "peer effect," and "regulations." The result analysis showed up benefits of "peer effect". So, it also demonstrated that rules have not much impact. The co-relation analysis was also performed over these effects and the outcome shows that almost all contestants are affected by peers on utilizing e-learning.

Luz Andrea Rodríguez Rojas et al. [43] discussed that health tourism is a popular sector in many countries in which industrialized countries are delivering improved, cheaper medical care. The above model is used for the management of expertise and is also focused on information and communication technology (ICT) in the medical tourism field. This model presented a tool for achieving upgraded decision-making process of the business as it contributes to knowledge, value, and productivity of an expertise. This management model provides a roadmap for these goals by developing executing instruments.

Suhaila Samsur et al. [44] discussed about gaining accurate knowledge to create a secure electronic medical record (EMR) system for security analysis of shared information. Previously, a preliminary survey of interviews was done with a chosen group of HIS developers. The result analysis showed that four factors are needed to be considered for designing privacy preserving framework for hospital information system (HIS). These factors are such as, legislation, ethical, technology and cultural. With application of KM, more systematic rules can be retrieved to achieve security of an information. Additionally, it envisioned light for designing more secure HIS.

Anna Závodská et al. [45] analyzed the role of KM and their process along with performance. The designed architecture in this work showed up the importance of each level of knowledge information. Comparison of knowledge is also discussed about different firms. Marketing campaign information is given. The method involves three case studies, which collect and analyze facts. The study indicates that the framework of the value-added mechanism can be used as an estimation method for analysis of values in many phases and different business management methods are appropriate. Proposed modifications are made on the grounds of the questions discussed. The blue ocean strategy principles are the basis for these recommendations. Blue ocean technology is considered to be a better approach to knowledge processing and the advancement of the organizations.

Fernando J. Garrigos-Simon et al. [46] considered the developments and transformations faced by ICTs in the development of the social networks and Web 3.0 technology in the contemporary economic and social environment because of the effects of recent innovations on corporate efficiency. The inquiry illustrates the use of social networks and the growth of digital business models associated with them, which highlight the importance of group managers and crowdsource platforms. In light of recent progress, the report explores possible sources of competitive gains for businesses and describes the improvements they should bring to improve decision-making and use current situations. The paper analyzes the impact of social network management and Platform 3.0 technology promotion and discusses methods for strengthening competitive advantages.

Muhammad Waseem Muhammad Zia et al. [47] discussed about third generation of Internet, Web 3.0. It is used to create external data storage facilities to allow any kind of data to be exchanged across any network. This covers the hypothesis that in the third century the location and the computer sciences were associated with the library. Various advantages of modern web technologies, Web 3.0, have been established to develop applications for library information science.

A huge number of scientists, researchers, and so on are involved in Network 3.0. Semantic platform systems allow consistent knowledge from various domains to be combined to reach special solutions. Currently, semantic web applications are not as popular as other types of enterprise services, so their early adopters can be encouraged by defining their consumer abilities. Ilona Pawełoszek et al. [48] introduced semantic web software characteristics. Some indications have been provided of the strategic position of semantic pharmaceutical and e-commerce applications.

Rajiv et al. [49] discussed that Web 3.0 have assisted in web-based learning that assists people in reading and writing. Web 2.0 and Web 3.0 have created innovations with a much shorter period, which allow for simple web-based schooling and learning. The development and functionality of Web 3.0 are defined in these papers. There will also be talks about possible future innovations, trends, applications, and so on.

Ju-Young M. Kang et al. [50] focused on whether (1) (a) economic tenderness, (b) conformity with the m-Internet equipment, and (c) leisure patterns lead to conditional rewards, social self-concept profits, emotional advantages, and functional benefits.

Pranita Barpute et al. [51] discussed that e-commerce is one of the most popular industries today. We add, remove, change, and update items to suit the demands of dealers and end customers in online shopping. The search article would be easier to run for users accessing the online shopping structure provided the computer has the correct user interface. This will help various distributors grow and bind them to their brand online. For the proposed platform, the semantic web and Web 3.0 can even be used to help users and also distributors. The interfaces for both buyers and dealers are shown.

Daisy Wadhwa et al. [52] discussed that the standard of living is completely changed by technology. In the current world, the accessibility of technologies of web provide us the way to establish path for collaborative learning. Collaborative learning is trending among all varieties of learning. This study spots the benefits and aims of collaborative learning online. The implementing outline of Web 3.0 from Web 1.0 is available by this study. This study spots the aim and benefits of online collaborative learning. The implementing outline of Web 3.0 from Web 1.0 is also discussed in this study. This study also provides insights about tools of Web 3.0 that makes the learners to collaborate among them and to learn and teach simultaneously. Then a review is also presented for "Wiki-Learning" as collaborative learning platform.

Many organizations are offering to install KM that has become very popular, for enhancing the performance of their organization. Hayfa. Y. Abuaddous et al. [53] inspected the literature for showing the effects of management of knowledge and some of its practices on organizations performance. According to the observations, the efficiency of KM is identified the direct or indirect relation with organization's performance.

Testing is a dominant activity for incrementing the quality of product and original information support. KM connected with the process of testing increments the capacity in identification and defects removing, coupling the area information and technology in the implementation of the process are discussed by Darío Enrique Soto Duran et al. [54]. The dominant ingredient for any organization desiring competitive benefits is the efficient management of the knowledge. This study describes that KM is the primary characteristic of the performance of an organization and the dominant tool for the proper running of an organization. So the organization has to accept the full advantage of knowledge value for managing, creating, applying knowledge effectively, and sharing. Taking care of three main components – processes, technology, and people – is required for managing the knowledge efficiently. Consequently creating, managing, sharing and utilizing knowledge effectively is quite important for an organization to utilize the advantages of KM was discussed by Funmilola Olubunmi Omotay.

Bahman Aghamirian et al. [56] studied the relationship between organizational behaviour and customer's knowledge management in e-commerce and envisioned 8 factors such as: knowledge for the customers, knowledge from customers, knowledge about customers, recording and spreading customers' knowledge, quality of the services and products, reasonable prices and satisfaction of customers. People who were all proficient and experts of organizational market of Mellat Bank were selected and 210 proficient who are working in companies of Iranian insurance for the statistic society were selected.

Diana Elena Ranf et al. [57] demonstrated the needs and requirement for the structured framework to cluster and relief the information reuse and new knowledge production in managing projects. These necessities are fulfilled by KM and gives way for enhancing the practices of project management. This paper covers a notice of profits and rewards obtained by developers by installing and accepting the strategy of project management. A great role of the paper is to build a way to use the management of knowledge in project management. This begins from the four stages of the KM cycle also installed in project management.

# 8 Proposed framework

HRM is a department in all organizations that can see and handle all requirements and aspects of employees. There are several jobs that HRM handles such as employee hiring, their job performance analysis, and benefits appraisal. HRM department is dedicated to the welfare of employees in an organization. The existing HRM approach has the following problems:
- Management of organizational knowledge compatibility with Web 3.0.
- Communication overheads among groups and individuals.
- Inefficient documentation and proper sharing of information among employees.

While keeping these in mind, the proposed framework is designed to coordinate the entire HRM system globally. The proposed framework is designed for using a machine learning approach over the cloud. The framework is designed in the distributed architecture described below (Figure 6).

The following steps are performed for proposed architecture:
1. The data collection unit collects the data from employees regarding their complaints, performance, appraisal, bonus distribution, etc.
2. Then the collected data is stored at the cloud server from where other units can access data.
3. Then analysis unit integrated with machine learning will perform analysis according to the requirement.
4. The result of the analysis is also stored at the cloud server.
5. Finally, decision-making unit is integrated with a machine learning approach such as fuzzy logic to decide for employees and to coordinate to them through a cloud server.

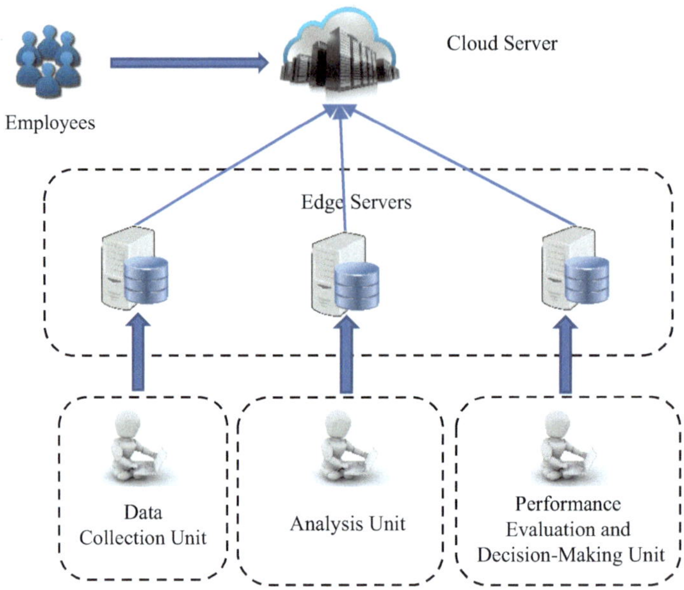


**Figure 6:** Proposed architecture for HRM integrated with ML over cloud.

Key features of the proposed architecture are:

1. In this research work, machine learning/AI is used to collect, analyze, and decisions making for employees. This will show the integration of Web 3.0 features (AI and cloud computing) for making advancement in HRM procedure.
2. KM or data mining approach is used for analytical purposes.
3. This model is quite efficient for finding the best solution for problems existing in the organization.
4. This decision support system also efficient with respect to time.
5. This model gives decisions without any partiality and indirectly helps in the development of the organization.

# 9 Conclusion

In this chapter, the impact of the KM evaluation with Web 3.0 usability for organizational knowledge generation, sharing, organizing, and their leverage in HRM is discussed. In this chapter, by using the advantages of Web 3.0 tools and techniques such as AI and cloud computing, the proposed framework is designed to make the smart HRM system supporting collaborative learning in employees' growth in an organization. The proposed framework will prove its feasibility of KM concerning organizational growth. As part of the continuity of this work, the experimental implementation of smart HRM is planned to enhance their efficiency level.

# References


[1] Helmke, S., Dangelmaier, W. & Uebel, M.F. (2002). CRM-systems as technology enabler for a customer-oriented knowledge management. 249. https://doi.org/10.1109/picmet.2001.952143

[2] Kong, F. & Cai, L. (2008). Customer knowledge management and research in E-commerce environment. Proceedings – 2008 International Symposium on Knowledge Acquisition and Modeling, KAM 2008, 231–235. https://doi.org/10.1109/KAM.2008.149

[3] Lian, G.G. & Hu, L. (2008). The research on knowledge management based on supply chain. 2008 IEEE International Symposium on Knowledge Acquisition and Modeling Workshop Proceedings, KAM 2008, 1187–1190. https://doi.org/10.1109/KAMW.2008.4810708

[4] Gunawan, A. & Kurnia, S.G. (2019). Knowledge management to increase the human resource quality case study at PT. GAI. Proceedings of 2019 International Conference on Information Management and Technology, ICIMTech 2019, 482–485. https://doi.org/10.1109/ICIMTech.2019.8843815

[5] Carden, L., Kovach, J.V. & Flores, M. (2020). Enhancing human resource management in process improvement projects. In: Organizational Dynamics, Elsevier Ltd, 100776. https://doi.org/10.1016/j.orgdyn.2020.100776.

[6] Oehlhorn, C.E., Maier, C., Laumer, S. & Weitzel, T. (2020). Human resource management and its impact on strategic business-IT alignment: A literature review and avenues for future research. The Journal of Strategic Information Systems, 29(4), 101641. https://doi.org/10.1016/j.jsis.2020.101641.

[7] Czarnowski, I. & Pszczólkowski, P. (2020). A novel framework for decision support system in human resource management. Procedia Computer Science, 176, 1548–1556. https://doi.org/10.1016/j.procs.2020.09.166.

[8] Jantan, H., Hamdan, A.R. & Othman, Z.A. (2010). Intelligent techniques for decision support system in human resource management. In: Devlin, G. (ed), Advances in Decision Support Systems, 342.

[9] Zaouga, W., Rabai, L.B.A. & Alalyani, W.R. (2019). Towards an ontology based-approach for human resource management. Procedia Computer Science, 151, 417–424. https://doi.org/10.1016/j.procs.2019.04.057.

[10] Ahammad, M.F., Glaister, K.W. & Gomes, E. (2020). Strategic agility and human resource management. Human Resource Management Review, 30(1), 100700. https://doi.org/10.1016/j.hrmr.2019.100700.

[11] Davenport, T. & Marchand, D. (2000). Is KM just good information management. In: Financial Times Mastering Information Management: Complete MBA Companion In Information Management, Great Britain, Pearson Education Limited.

[12] Davenport, T. & Prusak, L. (2000). 'Working Knowledge: How Organisations Manage What They Know', United States of America, Harvard Business School Press.

[13] Koulopoulos, T. & Frappaolo, C. (1999). Smart Things To Know About Knowledge Management, United Kingdom, Capstone Publishing Limited.

[14] Nonaka, I. & Takeuchi, H. (1995). The Knowledge Creating Company, New York, Oxford University Press Inc.

[15] Stewart, T.A. (1999). Intellectual Capital: The New Wealth of Organisations, London, Nicholas Brealey Publishing.
[16] (PDF) Overview on Knowledge Management. (n.d.). Retrieved December 30, 2020, from https://www.researchgate.net/publication/324603140_Overview_on_Knowledge_Management
[17] Ammar-Khodja, S. & Bernard, A. (2008). An overview on knowledge management. In: Methods and Tools for Effective Knowledge Life-Cycle-Management, Berlin Heidelberg, Springer. https://doi.org/10.1007/978-3-540-78431-9_1.
[18] Kurilovas, E., Kubilinskiene, S. & Dagiene, V. (2014). Web 3.0 – Based personalisation of learning objects in virtual learning environments. Computers in Human Behavior, 30, 654–662. https://doi.org/10.1016/j.chb.2013.07.039.
[19] Newman, R., Chang, V., Walters, R.J. & Wills, G.B. (2016). Web 2.0 – The past and the future. International Journal of Information Management, 36(4), 591–598. https://doi.org/10.1016/j.ijinfomgt.2016.03.010.
[20] Cheung, K.H., Yip, K.Y., Townsend, J.P. & Scotch, M. (2008). HCLS 2.0/3.0: Health care and life sciences data mashup using Web 2.0/3.0. Journal of Biomedical Informatics, 41(5), 694–705. Academic Press. https://doi.org/10.1016/j.jbi.2008.04.001.
[21] Nath, K., Dhar, S. & Basishtha, S. (2014). Web 1.0 to Web 3.0 – Evolution of the Web and its various challenges. ICROIT 2014 – Proceedings of the 2014 International Conference on Reliability, Optimization and Information Technology, 86–89. https://doi.org/10.1109/ICROIT.2014.6798297
[22] Alabdulwahhab, F.A. (2018), August 20. Web 3.0: The decentralized web blockchain networks and protocol innovation. 1st International Conference on Computer Applications and Information Security, ICCAIS 2018. https://doi.org/10.1109/CAIS.2018.8441990
[23] Ifenthaler, D. (2013). Introduction: Is Web 3.0 changing learning and instruction?. In: Towards Learning and Instruction in Web 3.0: Advances in Cognitive and Educational Psychology, New York, Springer. https://doi.org/10.1007/978-1-4614-1539-8.
[24] Azeem, M.F. & Yasmin, R. (2016). HR 2.0: linking Web 2.0 and HRM functions. Journal of Organizational Change Management, 29(5), 686–712. https://doi.org/10.1108/JOCM-09-2015-0152.
[25] Web 3.0. (2014). Encyclopedia of Social Network Analysis and Mining, New York, Springer, 2337–2337. https://doi.org/10.1007/978-1-4614-6170-8_110068.
[26] Dmitriy, C. What is Web 3.0. Version and evaluation. / D. Christophe – [Available at]: http://therunet.com/articles/1552-chto-takoe-web-3-0-versii-i-otsenki
[27] Human Resources System in Russia in 2013 HRM. Short version of the analytical report. [Available at]: – Mode of access: http://www.tadviser.ru/index.php/HRM
[28] Salacanis, J. Web 3.0, the "official" definition. / Jason Salacanis – [Available at]: http://calacanis.com/2007/10/03/web-3-0-the-official-definition
[29] Applebaum, E., Bailey, T., Berg, P. & Kalleberg, A.L. (2000). Manufacturing Advantage: Why High-Performance Work Systems Pay Off, Ithaca, NY, Cornell University Press.
[30] Barney, J.B. (1991). Firm resources and sustained competitive advantage. Journal of Management, 17, 99–120.
[31] Berggren, C., Bergek, A., Bengtsson, L., Hobday, M. & Söderlund, J. (2013). Knowledge Integration and Innovation: Critical Challenges Facing International Technology-Based Firms, UK, Oxford University Press.
[32] Rupietta, C. & Backes-Gellner, U. (2017). Combining knowledge stock and knowledge flow to generate superior incremental innovation performance – Evidence from Swiss manufacturing. Journal of Business Research.

[33] Malik, A., Froese, F.J. & Sharma, P. (2020). Role of HRM in knowledge integration: Towards a conceptual framework. Journal of Business Research, 109, 524–535. https://doi.org/10.1016/j.jbusres.2019.01.029.

[34] Jomthanachai, S., Rattanamanee, W., Sinthavalai, R. & Wong, W.P. (2020). The application of genetic algorithm and data analytics for total resource management at the firm level. Resources, Conservation and Recycling, 161, 104985. https://doi.org/10.1016/j.resconrec.2020.104985.

[35] Sołtysik-Piorunkiewicz, A. (2015). Knowledge management impact of information technology Web 2.0/3.0. the case study of agent software technology usability in knowledge management system. AIP Conference Proceedings, 1644(1), 219–227. https://doi.org/10.1063/1.4907840.

[36] Dominic, M., FrancisFrancis, S. & Pilomenraj, A. (2014). E-Learning in Web 3.0. International Journal of Modern Education and Computer Science, 6(2), 8–14. https://doi.org/10.5815/ijmecs.2014.02.02.

[37] El Hadi, M.M. & Amer, A.M. (2015). Educational knowledge management systems model based on Web 3.0 technology. Compunet (The Egyptian Information Journal), 14(16), 5–11. https://doi.org/10.12816/0025769.

[38] Sheth, A. & Thirunarayan, K. (2012). Semantics empowered Web 3.0: managing enterprise, social, sensor, and cloud-based data and services for advanced applications. Synthesis Lectures on Data Management, 4(6), 1–175. https://doi.org/10.2200/s00433ed1v01y201207dtm031.

[39] Morris, R.D. (2011). Web 3.0: Implications for online learning. TechTrends, 55(1), 42–46. https://doi.org/10.1007/s11528-011-0469-9.

[40] García, V.H.M., Carrillo, T.L. & Castro, S.J.B. (2013). Knowledge management model applied to a complex system: Development of software factories. Advances in Intelligent Systems and Computing, 172, AISC, 185–195. https://doi.org/10.1007/978-3-642-30867-3_17.

[41] Rosa Yeh, C. & Singhateh, B. (2013). The effect of connectivism practices on organizational learning in Taiwan's computer industry. Advances in Intelligent Systems and Computing, 172, AISC, 219–229. https://doi.org/10.1007/978-3-642-30867-3_20.

[42] Chang, T.K., Huang, H.F. & Chang, S.M. (2013). Understanding educational administrators' subjective norms on their use intention toward on-line learning. Advances in Intelligent Systems and Computing, 172, AISC, 267–273. https://doi.org/10.1007/978-3-642-30867-3_24.

[43] Rojas, L.A.R., Bermudez, G.M.T. & Morales, A.A.O. (2013). A knowledge management model applied to health tourism in Colombia. Advances in Intelligent Systems and Computing, 172, AISC, 537–546. https://doi.org/10.1007/978-3-642-30867-3_48.

[44] Samsuri, S., Ismail, Z. & Ahmad, R. (2013). Adopting a knowledge management concept in securing the privacy of electronic medical record systems. Advances in Intelligent Systems and Computing, 172, AISC, 547–558. https://doi.org/10.1007/978-3-642-30867-3_49.

[45] Závodská, A., Šramová, V. & Aho, A.M. (2013). The role of knowledge in the value creation process and its impact on marketing strategy. Advances in Intelligent Systems and Computing, 172, AISC, 589–600. https://doi.org/10.1007/978-3-642-30867-3_53.

[46] Garrigos-Simon, F.J., Alcamí, R.L. & Ribera, T.B. (2012). Social networks and Web 3.0: Their impact on the management and marketing of organizations. Management Decision, 50(10), 1880–1890. https://doi.org/10.1108/00251741211279657.

[47] Waseem Muhammad Zia, M. & Ahmed, F. (2019). Possible uses of web 3.0 in websites of Libraries of Academic Institutions of Pakistan. https://digitalcommons.unl.edu/libphilprac

[48] Pawełoszek, I. (2015). Web 3.0 applications in enterprise strategy. Studia Ekonomiczne., 234, 129–139.

[49] Lal, M. (2011). Web 3.0 in Education & Research. BVICAM's International Journal of Information Technology., 3(2).

[50] Kang, J.-Y.M. & Johnson, K.K.P. (2013). An Application of Web 3.0: Positive WOM for Mobile Location-Based Apparel Shopping Services Usage. International Textile and Apparel Association Annual Conference Proceedings, 70(1). https://iastatedigitalpress.com/itaa/article/id/2185/.

[51] Barpute, P., Solankar, D., Hanwate, R., Kurwade, P. & Avhad, K. (2016). IJARCCE building ecommerce framework for online shopping using semantic web and web 3.0. International Journal of Advanced Research in Computer and Communication Engineering, 5, https://doi.org/10.17148/IJARCCE.2016.53242.

[52] Wadhwa, D. (2015). A study of web 3.0 technologies for collaborative learning. International Journal of Engineering Technology, Management and Applied Sciences, 3, 92–97.

[53] Payal, R., Ahmed, S. & Debnath, R.M. (2019). Impact of knowledge management on organizational performance: An application of structural equation modeling. VINE Journal of Information and Knowledge Management Systems, 49(4), 510–530. https://doi.org/10.1108/VJIKMS-07-2018-0063.

[54] Dario Enrique Soto, D., Jovani Alberto Jiménez, B. & Adriana Xiomara Reyes, G. (n.d.). Application of Knowledge Management to the software testing process Aplicación de la gestión del conocimiento al proceso de pruebas de software (Vol. 38).

[55] Omotayo, F. (2015). Knowledge management as an important tool in organisational management: a review of literature. Library Philosophy and Practice (e-Journal). https://digitalcommons.unl.edu/libphilprac/1238.

[56] Aghamirian, B., Dorri, B. & Aghamirian, B. (2015). Customer knowledge management application in gaining organization's competitive advantage in electronic commerce. Journal of Theoretical and Applied Electronic Commerce Research, 10(1), 63–78. https://doi.org/10.4067/S0718-18762015000100006.

[57] Ranf, D.E. & Herman, R. (2019). Knowledge management contributions in project management. Land Forces Academy Review, 23(4), 288–293. https://doi.org/10.2478/raft-2018-0036.

Sandhya Java, Hiba Mohammed, Aradhana Balodi Bhardwaj, Vinod Kumar Shukla

# Education 4.0 and Web 3.0 Applications in Enhancing Learning Management System: Post-Lockdown Analysis in COVID-19 Pandemic

**Abstract:** Education 4.0 is a new vision for learning that enables educators and apprentices to align technology with research to discover new possibilities. The advancements made through the Internet, artificial intelligence, and virtual classrooms have drastically transformed the education industry. The onset of novel coronavirus forced the whole world into immediate lockdown, with schools and universities shifting to digital platforms to keep the learning process going. The pandemic affected about 90% of the student population globally and forced educational institutions to consider unconventional approaches for teaching. The modifications made to the existing system for the pandemic depended on resources derived from Education 4.0 which could be permanent. Many of these were technological upgrades to the current mode, focused more on student's development and learning from the initial stages, along with plans for more differentiated and inclusive lessons. However, the changes made currently or in the future would be long term which could have its challenges and advantages and require extensive planning through research for smooth implementation. **Web 3.0** assets derived from data integration have improved e-learning opportunities for students through personalization, knowledge management, and enhanced information delivery while providing assistance for instructors in course development, record keeping, and teaching support. Initially, this research investigates what are the new tools and technology being applied to the current structure. This chapter aims to examine the impact of these practices introduced as part of Education 4.0 on learning, post-COVID-19 for students and teachers. Highlighting the updates in Education 4.0 from previous revolutions has been strongly considered in this study due to the lack of awareness.

The chapter carefully analyzes existing literature on the trends of Education 4.0 extracted from the Industrial Revolution (IR) 4.0 initiated in 2012. A review was conducted to understand the correlation between IR 4.0 and **Web 3.0** practices being used for teaching students at different levels during the outbreak. An empirical-analytical methodology was undertaken to gather data through an online questionnaire designed for school and university students. The undertaken sample was questioned about their experience with e-learning tools during the pandemic and suggestions were taken for future education mode preferences. Through this

https://doi.org/10.1515/9783110722789-006

investigation, researchers could answer the impending question, can initiating Education 4.0 during this pandemic revolutionize the education system post-COVID and equip students with the skills needed to survive in the twenty-first century. Organizations and the school system need to understand the impact of these modifications on all the members of the learning community from a future perspective. Post-pandemic, researchers could study ways to better implement the specific tools discussed in this chapter and evaluate their effectiveness in different offline classroom environments. The shift to e-learning could be the much-needed change to the global education structure for a more inclusive and student-centered learning environment.

**Keywords:** Education 4.0, COVID-19, E-Learning, Technology, Web 3.0, Post-pandemic, Education System

# 1 Introduction

At present, much of what is known to us as the physical world has evolved through big advancements, and developments in different industries, pushing mankind to work more, better and faster. The technological changes in the last few decades have pushed us into an era of digitization with virtual assistants make our lives easier, 3D environments giving experiences like never before and netizens connecting with each other globally. All of the good and bad were possible only due to man's belief in proving itself as the superior race through revolutions of the industry.

Before the development and introduction of industry 4.0, there were three distinct industrial reforms that changed mankind's experience of the working environment and paved the way for greater technological shifts in the coming years. The first industrial revolution focused on work through manual labor with aids from animals and basic machine tools. Next leap was the second industrial revolution in the early 1920s with introduction of steel and electricity-powered factories which drastically increased productivity. With the emergence of electronic equipment, the third industrial revolution was pioneered with manufacturers adopting better digital technology and software for their workplace. The growth of Internet and computer technology in industry 3.0 formed the basis of communication and collaboration initiated in the fourth industrial revolution. With access to real time data, cyber physical systems and artificial intelligence the industry has unlocked an advanced level of interconnectivity which empowers businesses to understand and control their operations better than ever before while driving their efficiency [1].

The Industry Revolution 4.0 has been characterized as the fusion of the biological, physical and digital worlds with increasing application of newer technologies like, robotics, big data, 3D printing, artificial intelligence, and Internet of Things (IoT). These advancements have started a process of digitization of not only the business industry but consumers, education, travel, health, and economy as well especially during the COVID-19 pandemic. The fourth Healthcare revolution derived from Industry 4.0 has introduced newer digital tools to automize the health system and prioritize patient care. Like, AI systems assisted hospitals in keeping track of electronic patient data and medical supplies which reduced the staff workload, while Big Data models provided trends and infographics regarding the patients and hospitals in various countries during the pandemic [2]. Similarly, Industry 4.0 also initiated the digital revolution of the education sector known as, "Education 4.0" with learner-focused objectives and technological provisions for every stakeholder (student, teacher, and administration).

Education 1.0 was a system where memory and ample experience were much required with learners being considered to be the receptacles of knowledge. Then, came the Internet with the revolution where pupils broke the tradition and consumed knowledge for their daily life applications which was referred to as the education 2.0. Here, the apprentices were said to be communicating, collaborating, and co-creating. From there, technologies began more rapid movement. Students started to venture into new types of technologies. Reports suggested that learning should be experienced through collaboration and students should work in the team for production of good learning outcomes and this system was coined to be education 3.0. Learners were considered to be connectors, creators, and constructivists.

The very next and recent approach was Education 4.0 which aligned with the industrial revolution 4.0. This model contradicted the previous ideas as knowledge is not just for consumption and production, but innovating diversified ideas with smart technologies, artificial intelligence, and robotics, all of which now deeply impact our everyday lives. Education 4.0 could be called the general upward trend blending the new and existing tools to enhance the education system and prepare students for a future workforce of Industry 4.0 [3]. This system of learning incorporates different technologies and approaches to enable higher educational institutions to align their services and syllabuses to prepare the future graduates for a new advanced beginning.

As the human beings embrace the tech future, how and what we teach in our education system will undergo reshaping, to align the system and keep up to date with the technologies of the twenty-first century. The future of understanding and teaching has taken a great leap from the traditional methods to the novel methods as students give more time and creativity to their class performance using the resources given by the educators while needing lesser guidance. The concept of teachers instructing a room full of students who listen and respond to directions has long become the past of education. While not a complete approach, student learning spaces powered by digital resources will supersede the typical classrooms of today, creating a more personalized learning experience for the students [4], thus aiding students to become partners or co-creators of their own education (shown in Figure 1).

Facilitation of experiences that allow collaboration, communication, and teamwork for students needs to be enhanced and strongly considered even outside the classroom.

**Figure 1:** Trends of Education 4.0.

The students shall be allowed to interact in coherence among themselves while working on their own unique learning projects with groups larger or smaller. Classrooms will co-exist as physical spaces and virtual environments while the concept of flipped learning would be strongly considered, so that students can understand theory at home but spend in-class time working as teams to apply their knowledge to real life issues and discussions. The whole idea of learning anytime, anywhere has become crucial due to the current situation and should considered as the "new-normal" due to the pandemic, where social distancing is essential, and schools/universities are virtual. Providing free digital resources and virtual learning environments globally has been an important objective of Education 4.0. The new tech-driven educational systems of **Education 4.0** also gives an insight for teachers helping them produce individualized lessons for students according to their pace and provide interactive content most beneficial for them. This prepares the learners to become much more innovative, better data collectors, and seize any opportunities that are shared with them [5].

As we ride the wave of digital era, it has become essential to connect with people globally without any barriers. A world of information is at our fingertips, with a click of a button or a voice command. When it comes to digitization, it is crucial that students are provided with new, updated educational systems to cope with being in the world of web. Technology can enhance learning and should be embraced and manipulated for mankind's advantage. The chapter gives a detailed idea of certain technologies revolutionizing the education industry and how they can be applied for the benefit of various stakeholders in a school or university. Furthermore, a qualitative research design is employed to develop a better understanding of student's perspective on the Education 4.0 tools and technology during the pandemic.

## 2 Internet of Things (IoT)

IoT is simply the concept of connecting any device that collect, share, and act on information on surrounding the Internet. It covers a wide usage from our home to office apps which include smart and wearable devices and almost anything else that are connected to Internet and communicate with a common language; these objects have built in censors and they come in various shapes and sizes that receive information from the environment and connects to an IoT platform. These platforms integrate and analyze data.

The global population continues to grow and age as people live longer. This calls for innovative ideas to manage the planets resources and to deliver services to everyone. As people shift toward cities the challenges to manage traffic, pollution, and energy increase. With all these going on, individuals have access to information products and global services which are increasing day by day and they expect these to be tailored to their individual needs. Technology is believed to satisfy these needs and contribute to our vision of an augmented life for everyone. Different skills develop solutions to make the world smarter and more connected. Talking about IoT, it is simply the concept of connecting any device that collect, share, and act on information surrounding the Internet. It covers a wide usage from our home to office applications which include smart and wearable devices and almost anything else that are connected to Internet and communicate with a common language [6].

The recent editions of wireless technologies such as embedded sensors and actuator nodes, which gives the opportunity for IoT to transforming the Internet into a fully integrated future network, that receive information from the environment and connects to an IoT platform. The platforms integrate and analyze data and help provide various information about the world around us.

IoT is the term for non-traditional computing devices which are connected to the Internet in order to send data and receive instructions without human interventions. The universal recognition enabled by wireless sensor network (WSN) technologies come across different aspects of our modern-day lifestyles. This helps us in measuring and coming up with an inference which also gives us the opportunity to understand the environmental indicators from delicate ecologies and natural resources to urban environments [7]. The propagation of these high-tech devices in activation of communication network creates the IoT. It influences our daily routine and lifestyle; for example at homes, these smart techniques enable us to allow management of lights, heating, or any home appliance that embeds a smart system, basically giving us an opportunity to use smart home system where we can remotely monitor and control our home. Similar examples of IoT-based technologies would be smart watches that track your daily routine (steps, heart beats, etc.), smart cars, and smart phones. The data securities and secure communications ensure that your personal information always will be safe. Data is emitted from various sensors and IoT platforms receive it and integrate the collected data from various sources; further analytics are performed on the data and valuable information is extracted as per requirement. Finally, the result is shared with other devices for better user experience automation and improving efficiencies. Through these smart technologies IoT helps in redefining our lifestyle and transforming the way we interact with technologies [8].

## 2.1 5G technology with IoT

Fifth-generation wireless or what we currently call the (5G) is the latest reiteration of technology, concocted in order to increase the speed and response of networks, around the world. 5G is an invention which helps enable a sharp intensification in the amount of data transmitted over wireless systems due to more available bandwidth and advanced antenna technology. According to various reports 5G has a multiplied speed which is ten times faster than that of an LTE network. This great upsurge in the transference speed will allow IoT devices to communicate and share data faster. The development of 5G network system along with the IOT, which is considered to be applicable to a variety of promising fields and applications, helps in evolution of various cellular network technologies and that creates a possibility organize immense censors in the framework of IoT and also helps to process data, by challenging all the other technologies of communication or data mining [9].

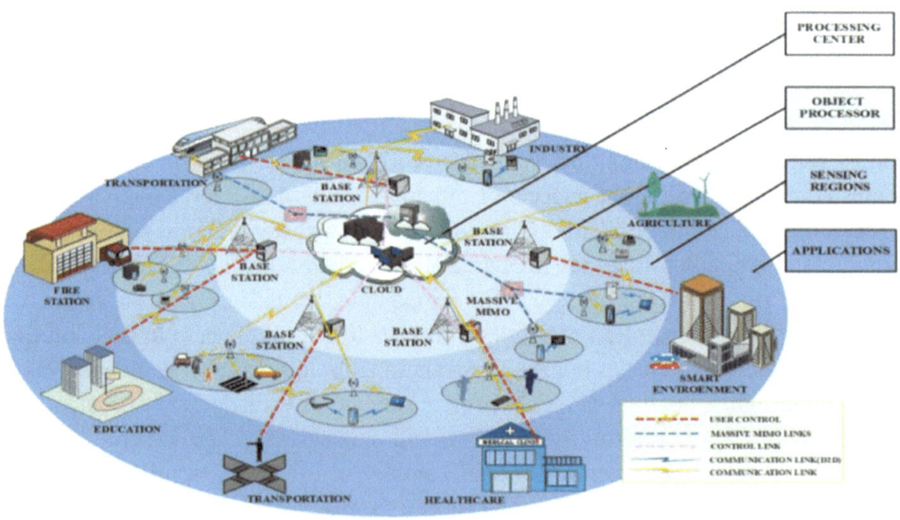


Figure 2: Development of 5G with IoT.

Figure 2 shows an expansion and conjunction of IoT with 5G network and that gives a perfect example of Internet, intelligence, and things. 5G users are no longer just normal smart phone handlers but they are also users of vehicular network, in which case the user interconnects these "vehicles" with others. Moreover, the increase in number of users and the various of the types of users, due to the interoperability with technologies like IoT, results in a diverse form of novel applications, a variety of subjective requirements and poses challenges in being able to understand and satisfy quality of experience (termed as QoE), which focuses on the entire service experience, across the board. Additionally, there are numerous other potential methods and technologies have been proposed, such as millimeter-wave (mm Wave), massive multiple-input multiple-output (MIMO), and machine-to-machine (M2M) communications, so that 5G networks can be well adapted to IoT. To sum up these smart techs help the world to connect with massive devices and make a great contribution to providing advanced and intelligent services for the ease of human beings [10].

## 2.2 Role of IoT in education

Throwing some light on the perspective of education, classrooms are not just confined to lecture halls. As mentioned in the following figure, especially during these times of pandemic, IoT has already made its way to campuses and schools throughout the world. There are thousands of sites that offer different courses, and you can access it from any part of the world. For millions of students with special needs IOT could provide benefits from educational resources that might not be available in classroom settings. Some of the examples are that there are mobile apps to help students with reading disabilities to read and understand materials, there are voice recognition apps that helps students with writing difficulties, and so on. The booming of IoT in education is adding more resources or tools to the classroom physically or online. Computers offer interactive audiovisual mediums and other software to present information in an effective and interesting way [11]. Through online platforms "study anywhere anytime" is a new form of phrase accepted by people all around the world especially during this pandemic time. No matter where you are in the world you just got to have Internet connection and a digital device, and a whole world of knowledge opens up to you. It is easily possible to learn a new skill or subject online. The novel corona time has made people use these opportunities to the maximum of their abilities.

Creating IoT products can be a great help to provide an easy environment for students, also reflected in Figure 3 [12]:

- Digital highlighters which scan any printed text and transfer it to your device making the note-taking process incredibly fast and easy.

- Radio frequency identification (RFID) chips can be used to track any physical object, for example,, plants and animals, in order to take up information and store them in the cloud for later analysis and inference.
- QR codes which can be embedded in books to access any additional resources or students can embed them in their offline work which can be linked to an online portfolio of the project.
- Smart classrooms – IoT will make classrooms of the future a tech haven for the teachers as well as students. Augmented reality (AR) helps students to give detailed information on any object they are studying on. For example, classes will be able to use textbooks encoded with AR possibilities. These types of techs have the ability to immerse students in a completely new environment. Students have the ability to explore anything up close from heavy machineries to human body and all the way to underwater ecosystem [13].

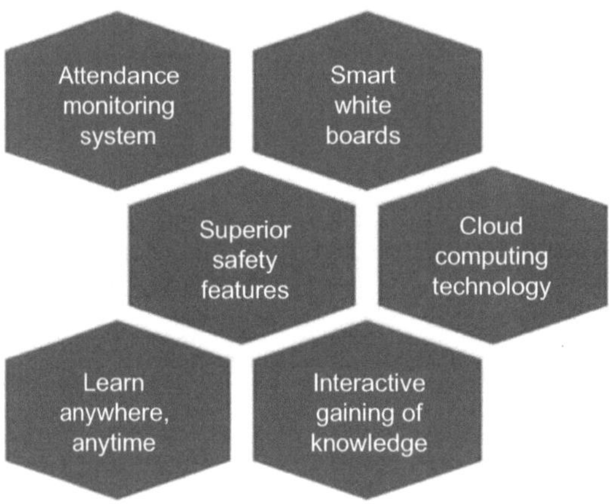


**Figure 3:** IoT education application.

Creating smart classrooms and different other new technologies for the purpose of educational and learning opportunities through multimedia and networking systems has helped teachers and the administration departments to interact and keep track of students individually or as a group, specifically during the lockdown period as it was hard to reach out to students without proper traditional classrooms. The content presented as optimization, interaction, and convenience of learning resources in smart classrooms is an advanced technical concept.

Cloud computing technology is a type of Internet-based computing. Server storage and applications are various services that are delivered over the Internet to the organization's computers and devices. Correspondingly 5G offers countless benefits to the management by offering opportunities for work and entertainment in the clouds to take advantage of cloud computing technology. This skill permits the reduction of the cost of computing application hosting, content storage, and delivery. Different cloud computing models are critical to getting the most out of the services offered by 5G intelligent classroom education systems. The ultra-reliable high-speed 5G waiting time is a catalyst for the application of this service [14].

Similarly, wearable device are the ones which help in students' issues by monitoring their movements in and out. These gadgets are connected to the management's central digital platform. Through these automations, it is not even necessary to take attendance of students as these devices automatically detect whether the student is present or absent. If at all the parents are worried regarding the whereabouts of their child, these technologies also might help them to check the student's location and their actions by sending them messages [15]. In combination with the IoT censors, these devices only allow the authorized people to access the door of the building and so when an unauthorized person arrives, the access is denied.

Likewise, further example is a report automation which is connected device that enables the school administrators to take advantage of AI to produce their weekly, monthly, or yearly reports on a wide range of issues, including student progress reports, financial planning reports, and reports for the school board and more.

## 2.3 Future of IoT

The future of IoT industry is seen as a huge success. According to *Business Insider* intelligence estimation 24 billion devices following the path of IoT will be installed by 2020 and IoT revenue will reach around 357 billion in the upcoming years which also results in a lot of job opportunities in the IT industry. As we move from www (static pages web) to web2 (social networking web) to web3 (ubiquitous computing web), the need for data-on-demand using sophisticated intuitive queries increases significantly. With the rise of cloud-based technologies along with data mining and big data analytics, the future of IoT seems to be promising and with it, the applications it is going to provide [16].

IoT (Figure 4) is going to expand at a very rapid rate and with it there are going to be huge advancements in every field. We can leverage the power of IoT to implement smart learning environment that facilitates better learning and greater retention rates. This advancement in education can produce better individuals in terms of skills and knowledge.

**Figure 4:** Development of IoT.

# 3 Artificial intelligence

The technology and innovations of fourth industrial revolution have rapidly boomed and revolutionized the processes of this planet. Artificial Intelligence, or AI, has been the driving force behind the Industry 4.0. AI is the machine replication of human intelligence through application of algorithms and development of a computing environment. Designed by computer scientists, AI enhances and imitates human capabilities in the form of machine learning and performs functions similar to us like, planning, reasoning, learning, and problem-solving. Research surrounding AI is concerned with developing machines and codes to automate daily tasks which require certain intelligence along with presenting solutions to real-life problems. AI systems are constantly being developed and used in different fields of business, military, medicine and military [17].

In regular household applications, AI is being applied as digital assistants in our mobile phones and smart home devices through Siri and Alexa which work on the popular machine learning software. Much of people's interactions online in chat boxes are powered by AI software that recognizes customer queries and responds based on their knowledge. Search engines for Bing, Google, and so on are operated through algorithms created through AI systems; these examples demonstrate the depth of human interaction with AI machines which has drastically increased in the past few years. Much of the manufacturing and business processes also depend on smooth functioning of AI and its understanding of user interface [18].

Natural learning processing (NLP) and machine learning are the most employed forms of AI that are used for image recognition, deep learning and planning, content extraction and text generation. Speech component of AI is highly utilized for machine translation and speech to text/text to speech recognition. Big organizations deploy these AI systems in different areas of working, the highest for external communication (71%), followed by customer interaction (66%), and workforce management and security (60%). AI-powered industries are expected to gain 30% more annually compared to their non-AI competitors [19].

## 3.1 Current trends of AI

Experts suggest AI would dominate the future of the world, especially considering the growing trends of machine learning software and AI skills in developed and developing countries. *Forbes* reports AI skills would penetrate the United States 100%, China 92%, and India 84% in the next 10 years of their development (as shown in Figure 5). Similarly, AI leaders of global tech companies suggest the next five years will have a big impact on the machine learning/Artificial Intelligence models created for enhancing processes due to regulation of data transmission and privacy [20]. Latest research also reports about 86% businesses running on AI models are deriving benefits through enhanced customer experience due to better understanding of certain human requirements and 25% AI-adopted business are expecting to see an increase in user-base and revenue in coming years [21].

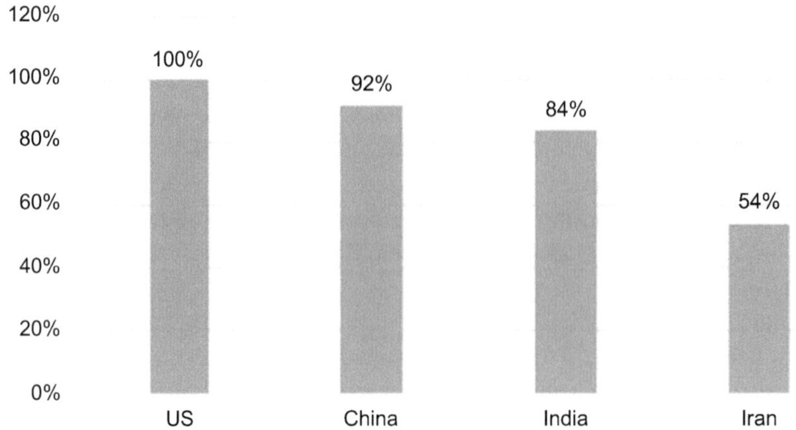


**Figure 5:** Prediction of artificial intelligence skills.

The 2020 pandemic has been especially important for the AI industry as the changes during this period have uncovered the true value of ML software through enhanced planning for workforce, customer retention, and simulation modelling. The growing trends of application of AI in augmenting and automating business processes in 2020 will continue as adopters of such technology report improvement of existing services and products.

In keeping with the transformations of Industry 4.0, many companies are also being forced to adopt AI-based solutions during this pandemic to address their business needs and obtain flexible solutions through advanced analytics and machine learning [22]. The education sector was deeply affected by the seismic shifts of the sudden worldwide lockdowns; all the stakeholders in a school/university/institute adopted applications (Zoom and Microsoft teams) with AI features to continue the learning process for students. Likewise, the pandemic has also introduced the need for emotion AI software equipped with the capability to understand complex human emotions and cognitive states through identifying vocal and facial expressions of users [23].

## 3.2 Role of AI in education industry

Artificial Intelligence and machine learning have long begun to alter the education industry, and the recent advancements in the technology have made it possible to increase access and personalize learning for the students. The various applications of the AI systems is changing the way students are learning and understanding, while making education easily accessible through computers and smart devices for all students especially for those unable to attend physical class. Such technology is impacting the learning process from playschool through advanced education, by offering better opportunities to enhance student experience with adaptive teaching features and personalized tools [24]. Other than learners, AI models are supporting teachers with managing tedious tasks and aiding education management bodies with automating administrative duties.

Specific software were designed during the Education 4.0 revolution through ML and AI modeling that have helped significantly improve a student's learning within and outside a classroom. Digital content can now be created by AI robots for classrooms; virtual interfaces can be developed using such robots for customized learning for students. AI system–powered application "Cram101" condenses contents in various textbooks and provides more digestible material via study guides, flashcards, and practice tests for students. AI technologies would now allow individualized learning to be prioritized by teachers by offering tools for differentiation in classrooms. Companies like Carnegie Learning are developing tutoring systems for students with specific needs and learning disabilities to offer lessons, tests, and instant feedback by using student provided data to identify knowledge gaps and create specific materials. Such applications are still in early testing stages, but soon could work as completely informed digital professors with broad array of learning styles to help students [25] (as displayed in Figure 6). Machine learning tech and predictive analytics are being explored to create and update content for textbooks, curriculum, and materials provided to students, in order to provide only current information and curate substance

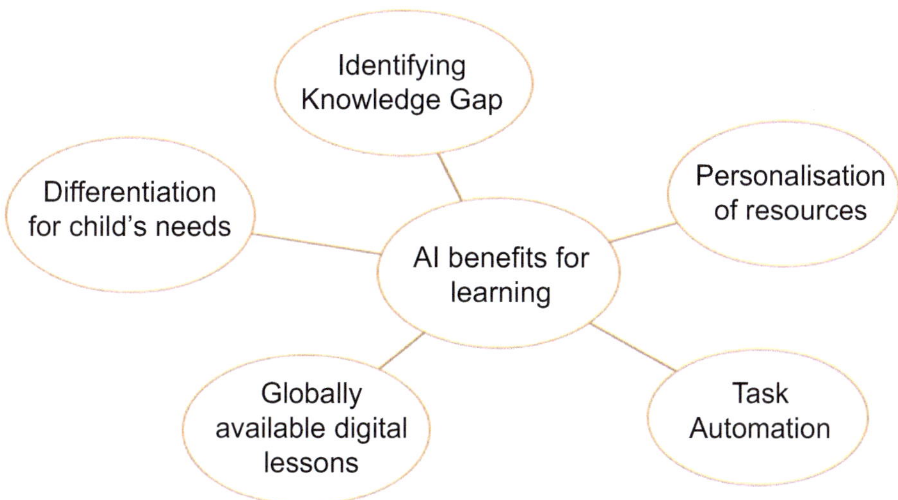


**Figure 6:** Benefits of AI for education.

for distinct learning curves. AI drives applications like Grammarly, Turnitin, and Kahoot to evaluate shared information and provide instant feedback to the learners. Immediate feedback given with resources to improve child's knowledge can increase their interest and motivation to study [26].

Virtual reality and augmented reality technology can be combined with AI to create modernized digital learning environment for students with stories, characters, and 3D gaming that make understanding a truly unique experience. Tools and online applications could create openly available, global classrooms for students with disabilities/physical impairments, or for those speaking different language. This opens up learning opportunities for children in remote parts of the world or those unable to attend classes due to illness like, during the pandemic [27]. AI could essentially bridge the gap between learning and accessibility of classrooms for learners.

For educators, AI has opened up a plethora of possibilities from reducing workload to improving class curriculums and lessons. One biggest job of teacher is grading and evaluating homework, while providing valuable feedback to students for improvement. AI applications have computerized the grading process for multiple-choice questions allowing teachers to spend personalized time with students instead [28]. Developers are also working on software to grade essays and written answers as well which could offer educators more time to provide adaptability and understanding during class.

Educations and lecturers have to design lessons and resources with differentiation for their students and content should be adapted to the abilities of twenty-first-century netizens. Platforms like Netex Learning permit teachers to create digital curriculum and content with online AI assistants through videos, audios, and other resources specifically designed for students with different needs [29]. During the pandemic, teachers were able to use many virtual AI platforms to increase student motivation in learning and improve their digital knowledge through customized training. By combining the best of teachers and technology, the goals of Education 4.0 can be achieved for best student outcomes.

Other than students and teachers, the management of schools and universities holds important responsibilities and administrative tasks that can be more efficiently accomplished with IT processes and virtual assistants powered by AI. Through modeling of previous data using ML, admin can keep track of departmental operations and obtain forecasts for better future planning. The forecasts could be used to record and maintain database of school supplies for students and food stock for cafeterias to avoid wastage in the future. AI significantly benefits the admission process through automating the classification of paperwork and enrolment process of student. This could also assist with eliminating the human bias related to admission process for fair and credible selection of future learners. AI systems can significantly improve the productivity of education institutes, while reducing the operation cost and increase their social visibility [30].

Researchers and scientists are exploring AI technologies to create smart virtual environments and digital platforms for students under Education 4.0. Reports suggest that the University of Southern California are developing robots that could play the role of a teacher or virtual educator by recognizing facial cues and gestures of students and respond with tailored verbal or non-verbal cues. Applications related to digital learning and simulation are projected as one of the highest benefit market in next few years. Research analysts for global AI have stated that AI tech in education industry will witness a 45% growth from 2018 to 2023 and the US education market is expected to reach 3.68 billion USD by 2023 (shown as Figure 7). Marketers also suggest that by 2025 about 47% of education tools would be developed using AI systems [31].

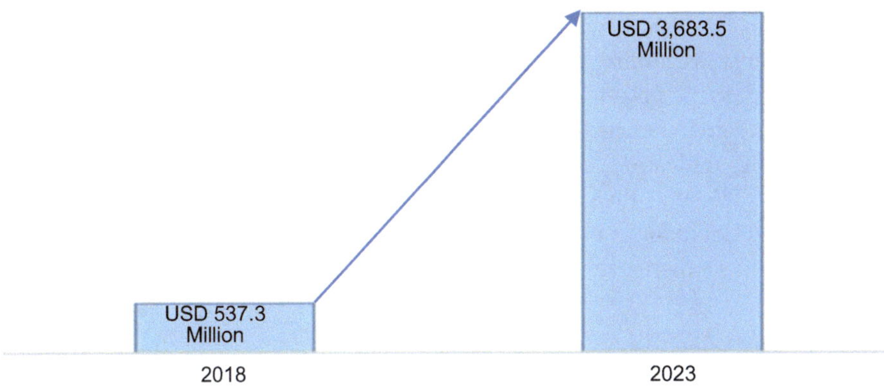


**Figure 7:** Expected growth of AI in education market.

Artificial Intelligence can competently drive personalization, efficiency, and globalization of education not just for students but for teachers and administration of education bodies as well. While the world witnesses an exponential growth of AI in education as well as other industries around the world, it is crucial for stakeholders to understand the benefits and invest in adopting AI technology to create a better learning environment for students, teachers, and the administration.

# 4 Methodology

Research methodology is the process used to identify the problem and select and analyze data collected from participants to accomplish the objectives discussed in the paper [32]. The data was gathered using survey data collection method. This has helped us gain a better understanding of student's experience with Education 4.0 technology. The sample for data collection consisted of 315 students (125 School and 190 University level) of mixed genders from India and Middle East. The sample was chosen on the preferred attributes using convenient sampling technique. A good rapport was established with the respondents and the instructions regarding the survey were clearly mentioned in the form. An online survey was developed with questions regarding Education 4.0 and its technologies for students, with general questions to identify their current level of education. Confidentiality of subject information was completely maintained. The data collected via the survey was analyzed and presented in the form of graphical representations.

# 5 Findings

A research survey regarding the application of Education 4.0 tools and technology during the pandemic for e-learning was developed and distributed to collect information on student's perspective. Analysis of data from 310 participants of schools and universities indicated a positive interest in students toward the introduction of these digital tools into their learning environment. The pandemic has surely affected the traditional forms of learning, yet proper application of technology has enhanced student's experience.

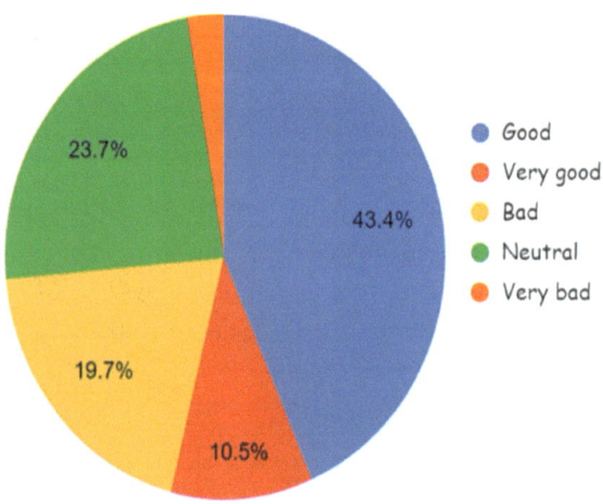


**Chart 1:** "How would you describe your online learning experience during the pandemic?".

For the research, the respondents were first asked about their general experience with digital classrooms during the pandemic. As shown in Chart 1, 43.4% subjects had a good experience, 23.7% had a neutral feeling, and 10.5 % had a very good experience with the shift on education onto digital platforms. Majority of the students believe the various online tools and technology used by educators during the lockdown were valuable and enhanced their experience.

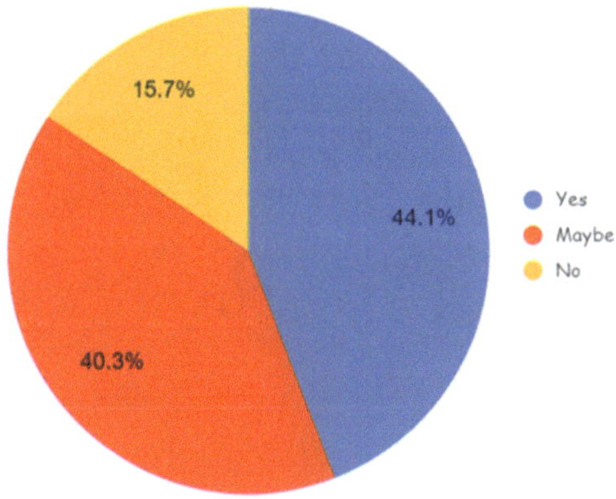


**Chart 2:** "Do you think the resources and tools shared by schools and universities during the pandemic have made smart digital classrooms, a better learning experience?".

Referring to Chart 2, we could analyze that 44.1% of the sample population agrees that the resources shared by schools and universities has made online education, an enhanced learning experience. A 15.7% of the sample claimed that the tools and resources of Education 4.0 were not useful or did not assist them to create a better learning experience for their digital classrooms during the pandemic. Yet there are 40.3% who had a neutral stand on the effectiveness resources.

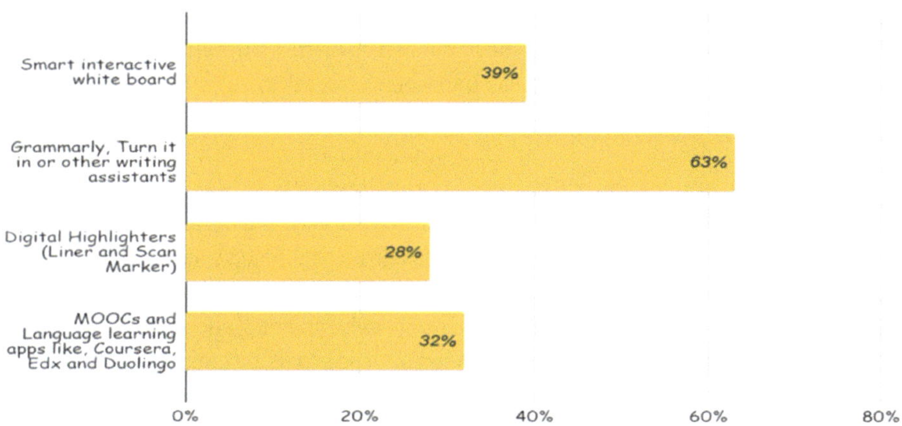


**Chart 3:** "Have you utilized any of these Education 4.0 tools for digital learning?".

From the research conducted in this chapter, certain technologies of AI and IoT were most highlighted for use in schools and universities, such specific Education 4.0 developed tools were given as options to the respondents to understand which of them were useful for their education. Results show, 63% students have used Turnitin, Grammarly, and other such AI writing assistants for their learning, 39% have been taught with or learnt using smart interactive white boards, 32% have used AI-driven MOOCs and language applications to improve their skills, and 28%

have used digital and smart highlighters to make study material easier to understand (shown in Chart 3). This indicates that a lot of IoT and AI tools developed for the education sector are being utilized by students to enhance their digital learning.

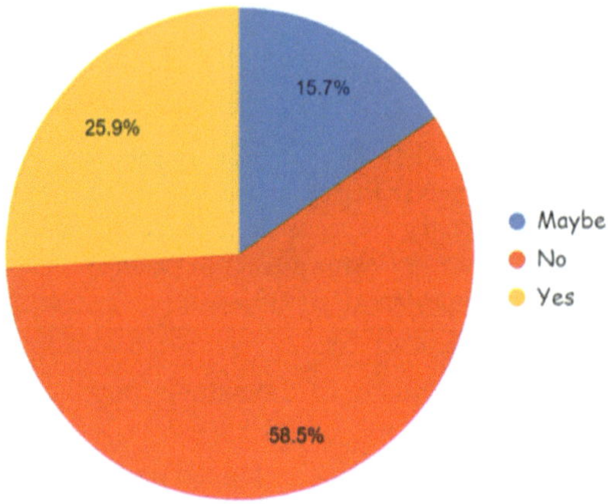


**Chart 4:** "Have applications of Web 3.0 (like Alexa, Siri, Virtual reality, and 3D graphics) assisted you in classrooms or during the education process?".

Referring to Chart 4, elaborates on the expediency of Web 3.0 and its applications with online classes or during the education process. As shown 58.5% of the population responds with "they have not" found tools of Web 3.0 useful, with 15.7% taking a neutral stand on the on the usefulness of Web 3.0 and its applications. Although there are 25.9% agreeing that web 3.0 had been applied in their learning environment and assisted in their online education. This explains that not much of the population is aware about Web 3.0 and its applications or the tools have not been properly put into application for student's education.

As students were inquired about their experience with digital tools for e-learning used in classrooms, it was crucial to understand from them suggestions to make their experience better as well. Majority of the responses (as shown in Chart 3) focused on making classes "more interactive with 3D/virtual environments" and providing "digestible content with variety of animation and media." The others recommended "initiating blended learning in classrooms" and "Personalisation of lessons and improve work feedback" to improve their learning experience in schools and universities. The AI and IoT systems discussed in this chapter could be appropriately applied through different mediums to achieve positive student outcomes.

After analyzing the sample data, the findings shows that the new technologies are still not considered due to lack of awareness of the mentioned applications by the sample population. Most of the students used the newest tools of Education 4.0 for utmost involvement and improvement in their digital learning experience. From the responses, we understand that proper implementation of e-learning resources during the pandemic has enhanced the classroom environment for students as virtual classrooms and increase in digital resources for lessons were given higher preference for

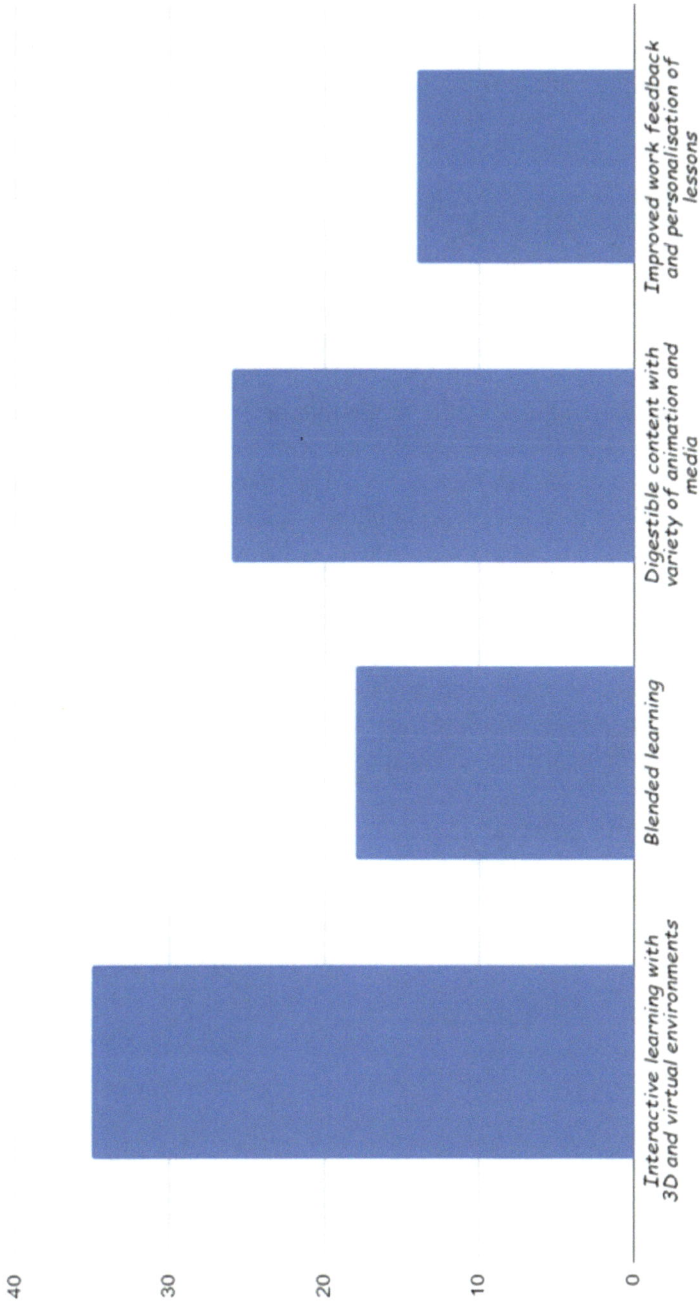


**Chart 5:** "If no, what do you suggest can make a student's experience better?".

post-pandemic situations. The recommendations given through this research should be strongly considered by educators and school/university management to improve the overall quality of learning provided to students.

# 6 Web 3.0

Web-based education and learning has opened up a vast world through the evolution of Internet and has given us the opportunity to access incredible prospects. As shown in the Figure 8, the evolution of web-based technologies started from the outmoded version of web which was coined as web 1.0 and established itself as "read-only" medium. This was the very birth of the Internet. It is like a centralized network wherein the main server hosts a website and with different servers getting access to this website. The main drawback of such a technology was that if at all one server went offline the entire website went offline. The next generation was of the Web 2.0 which it is said to a "read/write" medium. This was the creation of enhanced Internet services which are presently used by people globally, for example Facebook, Netflix, and Google. The Web 2.0 can be explained as distributed technological systems where servers are all around and controlled by multiple data centers, where each of them will maintain a copy of the data and thus the advantage is that even if one of the data centers is offline the rest of the team would still be functional and there is no gap in availability of the application [33].

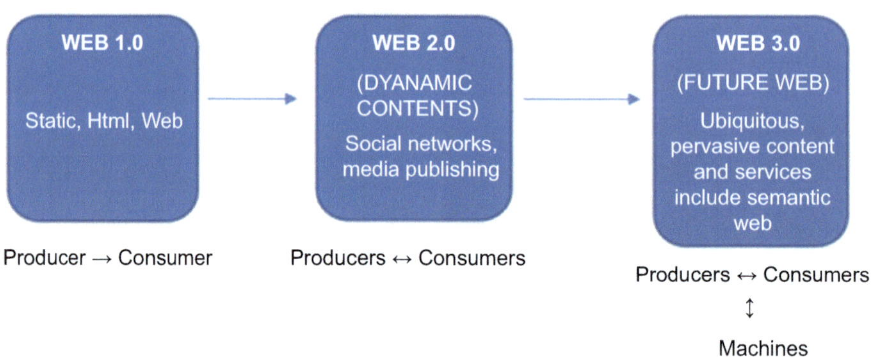


**Figure 8:** Development of Web 3.0.

The current most updated version of the web is coined as Web 3.0. This newer, advanced version has no ownership and helps in rebuilding internet as the original creators intended of the internet which means freedom, censorship resistant, anyone can participate and generate a sustainable income, more competition, and lower barrier and in the present scenario it has an easier entry to the marketplace. Web 3.0 consists of a decentralized system and every node of the server is able to maintain a copy of the data and moreover each node is able to validate all of the data which means faster validation of what any person has previously added to the implementation [34].

One of the best examples would be bitcoin where all of the data that is processed through the application will be time-stamped. This technology is an unconventional era of automation which allows users to not only read and write but also execute and innovate their own ideas. This mechanism has allowed machines to implement reasoning and problem solving which was only expected from the human brain. Within a span of short time, the Web 2.0 and the novel Web 3.0 technology is on the verge of creating new tools to simplify work and aid each individual for facilitating web-based education and learning [35].

The Web has endured extensive development and is undeniably growing faster. World Wide Web (www) has been used to improve our collaboration and communication, while making exchange of resources and delivering quality content easier. The world is in the third generation of the internet and specific innovations help us better define the features most useful for us. The semantic web has enhanced our ability to search and analyze the true meaning of words and data to help us generate, connect, and share content globally. Google, Facebook, Microsoft, and other big tech giants have adopted semantic web technologies to improve their database structure and search engine responses. Another enhanced feature of Web 3.0 is virtualization through 3D graphics and high-speed connection bandwidths [36]. This trend used three-dimensional applications to create virtual worlds/environments for exploration like the app Second Life.

With Web 3.0 our online search options and results have excelled as well. Search engines are no longer restricted to just text-inputs but also allow use of various media objects for search inputs. For example, Google allows us to add images for our device to retrieve other images with similar features as the objects in the input; similar technology has been developed to allow use of videos and audios as well. At this stage of the Internet, developers and researchers have made a remarkable progress with technologies for wireless networks and communication. The ubiquitous devices and platforms now allow us to access and manipulate a multitude of data from anywhere and anytime, even while being on the move, like Apple smart watches which allow us to control home appliances and act as our personal assistant. Such technology has improved our quality of life and the world's economy; with constant development, it is possible for us to reach greater possibilities virtually [37].

## 6.1 Applications of Web 3.0 technology in education

Over the last few years, different educational institutes have been offering online services through the web for admissions, management of activities and digital learning environments that connect students with excellent resources and each other for a better understanding. The services and tools of Web 3.0 have helped foster an open approach toward learning [38]. Features of technology could be carefully applied to the workings of the education sector to benefit students, teachers, and the management of schools/universities. The quality learning assets can be made advanced through ubiquitous devices and platforms, which could also be used to track the location and progress of students in their learning environment. The use of semantic web has a tremendous potential to enhance education for students, as our access to various interests and activities through our preferred channels is wide open. Educators could utilize semantic blogs and intelligent search engines to improve their education content with a variety of media and animation [39].

**Figure 9:** Technologies of Web 3.0 for education.

Similarly, a student's digital classroom can be improved using 3D technology and augmented reality to create virtual learning environments like, labs, museums, and zoos to make understanding and collaborating an enjoyable experience. Applications like Google Virtual Tour provide students the opportunity to explore parts of outside world that may not be easily accessible. Other such technologies further applied with AI and IoT tools can create better opportunities and environments for students.

# 7 Conclusion

This chapter explored the various trends of Education 4.0 extracted from Industrial Revolution (IR) 4.0. Systems and models reviewed for existing literature on IoT and

Artificial Intelligence have given us a greater understanding of its application in the education sector. Smart virtual assistants and school management systems powered by AI models could improve the quality of education provided. Also, applications like digital highlighters and smart white boards would enhance learning resources and methods of teaching. An open, personalized approach toward learning has been initiated through the advanced tools of Web 3.0. The findings gathered from the questionnaire designed for a sample of 315 students regarding their experience with Education 4.0 tools gave us an overall positive perception. Students were satisfied with their digital learning experience during the pandemic and offered suggestions for interactive and blended education. Even though learners preferred returning to traditional classrooms to continue knowledge development, their experience with technology has fostered a stronger need for educators and management to inculcate digital resources for their learning.

This chapter provides a solid foundation to explore other novel technologies of Education 4.0 which need to be reviewed and studied for its application in the current education industry. Through this study we would like to establish an emphasis on providing netizen students education on proper use of technology and resources with AI and IoT for flexible, improvised skills. Future recommendations for researchers and developers would be to carefully consider the application of technology for education especially for the post-pandemic situation or future scenarios. Providing students and teachers with easy access to multiple resources and secure technology could create an improved learning experience. Proper models of AI and IoT technology should be developed in collaboration with Web 3.0 tools to explore opportunities and possibilities of the fourth education revolutions. Advanced technology accurately applied with learning can open a Pandora of prospects for learners and educators.

# References

[1] Crnjac, M., Veža, I. & Banduka, N. (2017). From concept to the introduction of industry 4.0. International Journal of Industrial Engineering and Management (IJIEM), 8(1), 21–30.

[2] Jayaraman, P.P., Forkan, A.R.M., Morshed, A., Haghighi, P.D. & Kang, Y. (2019, Dec.). Healthcare 4.0: a review of frontiers in digital health. WIREs Data Mining and Knowledge Discovery, 10(2). doi: 10.1002/widm.1350.

[3] Bonfield, C.A., Salter, M., Longmuir, A., Benson, M. & Adachi, C. (2020, Jan.). Transformation or evolution?: Education 4.0, teaching and learning in the digital age. Higher Education Pedagogies, 5(1), 223–246. doi: 10.1080/23752696.2020.1816847.

[4] Keser, H. & Semerci, A. (2019, Aug.). Technology trends, Education 4.0 and beyond. Contemporary Educational Researches Journal, 9(3), 39–49. doi: 10.18844/cerj.v9i3.4269.

[5] Bonfield, C.A., Salter, M., Longmuir, A., Benson, M. & Adachi, C. (2020, Jan.). Transformation or evolution?: Education 4.0, teaching and learning in the digital age. Higher Education Pedagogies, 5(1), 223–246. doi: 10.1080/23752696.2020.1816847.

[6] Gaur, B., Shukla, V.K. & Verma, A., Strengthening people analytics through wearable IOT device for real-time data collection 2019 International Conference on Automation, Computational and Technology Management (ICACTM), London, United Kingdom, 2019, 555–560, doi:10.1109/ICACTM.2019.8776776.
[7] Allam, Z. & Dhunny, Z.A. (2019, Jun.). On big data, artificial intelligence and smart cities. Cities, 89, 80–91. doi: 10.1016/j.cities.2019.01.032.
[8] Shukla, V.K. & Singh, B., Conceptual framework of smart device for smart home management based on RFID and IoT 2019 Amity International Conference on Artificial Intelligence (AICAI), Dubai, United Arab Emirates, 2019, 787–791, doi:10.1109/AICAI.2019.8701301.
[9] "Unlocking IoT Data with 5G and AI", Innovate, Jun. 26, 2019.
[10] Wang, D., Chen, D., Song, B., Guizani, N., Yu, X. & Du, X. (2018, Oct.). From IoT to 5G I-IoT: The next generation IoT-Based intelligent algorithms and 5G technologies. IEEE Communications Magazine, 56(10), 114–120. doi: 10.1109/mcom.2018.1701310.
[11] Ajazmoharkan, Z., Choudhury, T., Gupta, S. & Raj, G., "Internet of Things and its applications in E-learning," 2017. [Online]. Available: https://fardapaper.ir/mohavaha/uploads/2017/10/Internet-of-Things-and-its-applications-in-E-learning.pdf.
[12] Alam, T. & Benaida, M. (2020, May). Blockchain and internet of things in higher education. Universal Journal of Educational Research, 8(5), 2164–2174. doi: 10.13189/ujer.2020.080556.
[13] Bai, H. & Zhang, Q. (2020). English smart classroom teaching system based on 5 network and internet of things. Microprocessors and Microsystems, 103421. doi: 10.1016/j.micpro.2020.103421.
[14] Banu, J.F., Revathi, R., Suganya, M. & Merlin, N.R.G. (2020, Jan.). IoT based Cloud Integrated Smart Classroom for smart and a sustainable Campus. Procedia Computer Science, 172, 77–81. doi: 10.1016/j.procs.2020.05.012.
[15] Ciolacu, M.I., Binder, L. & Popp, H. (2019). Enabling IoT in education 4.0 with biosensors from wearables and artificial intelligence. IEEE Xplore, Oct. 01, https://ieeexplore.ieee.org/abstract/document/8990763 accessed Jan. 07, 2021.
[16] Zantalis, F., Koulouras, G., Karabetsos, S. & Kandris, D. (2019, Apr.). A review of machine learning and IoT in smart transportation. Future Internet, 11(4), 94. doi: 10.3390/fi11040094.
[17] Vinuesa, R. et al. (2020, Jan). The role of artificial intelligence in achieving the Sustainable Development Goals. Nature Communications, 11(1). doi: 10.1038/s41467-019-14108-y.
[18] Marr, B. (2019, Dec. 16). The 10 best examples of how ai is already used in our everyday life. Forbes.
[19] Cockburn, I., Henderson, R. & Stern, S., "Nber Working Paper Series the Impact of Artificial Intelligence on Innovation," 2018. [Online]. Available: https://www.nber.org/system/files/working_papers/w24449/w24449.pdf
[20] Zhang, B. & Dafoe, A. (2019). Artificial intelligence: American attitudes and trends. SSRN Electronic Journal, doi: 10.2139/ssrn.3312874.
[21] Borana, J. "Applications of Artificial Intelligence & Associated Technologies," 2016. https://www.semanticscholar.org/paper/Applications-of-Artificial-Intelligence-%26-Borana/d5b061e6565ce421b4b0b7d56296e882085dc308?p2df (accessed Jan. 08, 2021).
[22] Bullock, J., Luccioni, A., Pham, K.H., Sin, C. & Luengo-Oroz, M. (2020). Mapping the Landscape of Artificial Intelligence Applications against COVID-19. arXiv.org, https://arxiv.org/abs/2003.11336, accessed Jan. 08, 2021.
[23] Chassignol, M., Khoroshavin, A., Klimova, A. & Bilyatdinova, A. (2018). Artificial intelligence trends in education: a narrative overview. Procedia Computer Science, 136, 16–24. doi: 10.1016/j.procs.2018.08.233.

[24] Deep learning goes to school: toward a relational understanding of AI in education. Learning, Media and Technology, 2020, https://www.tandfonline.com/doi/abs/10.1080/17439884.2020.1686017 accessed Jan. 08, 2021.
[25] Ciolacu, M., Tehrani, A.F., Beer, R. & Popp, H., Education 4.0 – Fostering student's performance with machine learning methods 2017 IEEE 23rd International Symposium for Design and Technology in Electronic Packaging (SIITME), Oct. 2017, doi:10.1109/siitme.2017.8259941
[26] Ciolacu, M., Tehrani, A.F., Binder, L. & Svasta, P.M., Education 4.0 – artificial intelligence assisted higher education: early recognition system with machine learning to support students' success 2018 IEEE 24th International Symposium for Design and Technology in Electronic Packaging (SIITME), Oct. 2018, doi:10.1109/siitme.2018.8599203.
[27] Yusof, A., Adnan, A., Mustafa, N. & Ahmad, M. (2019). Education 4.0 immersive learning with spherical videos (360°) and Virtual Reality (VR) experiences. ResearchGate, Nov. 05, https://www.researchgate.net/publication/337146653_Education_40_Immersive_Learning_with_Spherical_Videos_360_and_Virtual_Reality_VR_Experiences accessed Jan. 08, 2021.
[28] Azah Mansor, N., Abdullah, N. & Abd Rahman, H. (Jul. 2020). Towards electronic learning features in education 4.0 environment: literature study. Indonesian Journal of Electrical Engineering and Computer Science, 19(1), 442. doi: 10.11591/ijeecs.v19.i1.pp442-450.
[29] Shen, L. & Su, A. (2020). The changing roles of teachers with AI. Revolutionizing Education in the Age of AI and Machine Learning, 1–25. doi: 10.4018/978-1-5225-7793-5.ch001.
[30] Vazhayil, A., Shetty, R., Bhavani, R.R. & Akshay, N., Focusing on teacher education to introduce ai in schools: perspectives and illustrative findings 2019 IEEE Tenth International Conference on Technology for Education (T4E), Dec. 2019, doi:10.1109/t4e.2019.00021.
[31] Williamson, B. & Eynon, R. (2020). Historical threads, missing links, and future directions in AI in education. Learning Media and Technology, Jul., doi: 10.1080\/17439884.2020.1798995.
[32] Mohajan, H. (2018). Qualitative research methodology in social sciences and related subjects – munich personal RePEc archive. Uni-muenchen.de, Dec., doi: https://mpra.ub.uni-muenchen.de/85654/1/MPRA_paper_85654.pdf.
[33] Rajiv, & Lal, M. (2010). Web 3.0 in Education & Research. ResearchGate, https://www.researchgate.net/publication/284114717_Web_30_in_Education_Research, accessed Jan. 07, 2021.
[34] Rudman, R. & Bruwer, R. (2014). Defining Web 3.0: opportunities and challenges | Emerald Insight. The Electronic Library, doi: 10.1108\/EL.
[35] Barassi, V. & Treré, E. (2012, Jun.). Does Web 3.0 come after Web 2.0? Deconstructing theoretical assumptions through practice. New Media & Society, 14(8), 1269–1285. doi: 10.1177/1461444812445878.
[36] Miranda, P., Isaias, P. & Costa, C., "E-Learning and Web Generations: Towards Web 3.0 and E-Learning 3.0," doi:10.7763/IPEDR.2014.V81.15.
[37] Aljawarneh, S.A. (2019, Feb.). Reviewing and exploring innovative ubiquitous learning tools in higher education. Journal of Computing in Higher Education, 32(1), 57–73. doi: 10.1007/s12528-019-09207-0.
[38] Dominic, M., Francis, S. & Pilomenraj, A. (2014). E-learning in web 3.0. Modern Education and Computer Science, 2, 8–14. doi: 10.5815/ijmecs.2014.02.02.
[39] Ohei, K. & Brink, R., "Web 3.0 and Web 2.0 Technologies in Higher Educational Institute: Methodological Concept towards a Framework Development for Adoption," undefined, 2019.

Tapash Rudra, S. B. Goyal

# Artificial Intelligence (AI)-Aided Algorithms in Medical Practice with Special Emphasis on Genetic Disease Diagnostics

**Abstract:** The modern-day medical practice has been reaching greater heights every day with the latest technological improvisations. Medical fraternity is now capable of combating the deadliest of diseases with the advent of sophisticated tools and techniques. However, some sections of medical world remain obscure in regard to the diagnosis, prognosis, and therapeutics. Genetic or rare diseases are the crucial among them. Unlike the commonly known disorders, genetic diseases are quite unique primarily due to their association with the human genome. Moreover, scarcity of authentic data, restricted number of patients across the global periphery, and lack of adequate diagnostic measures are equally contributing to the hidden facets around such diseases. Therefore, it is imperative to have new and improvised options in order to find the unexplored facts related to them. Artificial intelligence and machine learning algorithms have appreciably enlightened various wings of rare disorders including the diagnosis. However, to make this new technology a powerful weapon, sizeable efforts are required bearing in mind genomic data and conventional methods cannot be the sole way out toward the improvement of genetic disease research and management.

**Keywords:** Artificial Intelligence, Algorithm, Machine Learning, Genetic Disease, Diagnostics

## 1 Overview

Human beings have been evolving since the first day on earth in tandem with the human civilization itself. The sustenance of humankind, however, continued to march on with the roller coaster ride. Our ancient ancestors were used to be intimidated by voracious carnivores and periodical natural calamities until coincidentally they managed to ignite the fire by generating the friction among the sharp edges of the stones. That was the beginning of the technology and it is beyond doubt that is the biggest boom we, human beings, have in the repertoire.

As the juggernaut moves on we find ourselves more toward the technology-driven era. The ongoing global COVID-19 pandemic has been humongous by far, in relevance to the overall collapse of the global integrity. To overhaul the outcome of such a colossal event we must come up with ideas those hitherto been thought about.

https://doi.org/10.1515/9783110722789-007

With the advent of new and sophisticated technologies medical science has been galloping to a new orbit. However, since the last few decades the implication of artificial intelligence (AI) has added a completely new dimension to it. The influence of various AI methods put a dilemma whether the natural or artificial intelligence is the potential roadmap to the global healthcare system. To be honest, diagnosis of diseases leading to the successive treatments is not replaceable but this modern boom of technology can enormously facilitate the healthcare professionals to ascertain the verdict of a disease. At the same time, the modern AI-directed analytics has opened up a new window to dissect the huge number of clinical data into relatively easier forms. This has made the task easier for the healthcare professionals and clinicians to be sure about the decision-making aspects [1–5].

## 1.1 Artificial intelligence in medical research: an account of current scenario

The holy grail of AI is the interpretation of human beings as the machines. To add to it, AI-based machineries are tailor-made to materialize the general theories of intelligence in agents. Biological entities such as humans and animals are treated beyond their limits in a wider perspective to maximize the outreach of the possible outcome. The notion of AI is extremely enterprising; at the same time, it is a force to be reckoned with to tackle the philosophical doubts those have been existing since more than thousand years.

We are heading more toward the Industrial Revolution 4.0 which is the narrative of fast-paced innovative development by securing greater efficacy in line with digitalized supply chain instead of the conventional one. Medical science is one of the potential thrust areas as far as the applications of Industry 4.0 are concerned. Improvised implants, techniques, and modern devices are some of the significant attributes. Moreover, Internet of Things (IoT) and Internet of Services has excelled the overall progression in relevance to the healthcare sector in the modern days [6]. Another striking aspect of Industry 4.0 is the replacement of conventional computing by means of quantum or cloud computing to propel the efficacy by several folds. This is very much noteworthy as in today's medical world the implication of quantum computing has reduced the cost to a considerable extent. Furthermore, use of sophisticated AI sensors has also escalated the productivity and efficiency among the various machines. The mentioned concept is a handful especially in strategizing the clinical practice [7, 8].

### 1.1.1 Significance of Industry 4.0 in medical fraternity

The principal features of Industry 4.0 that make it significantly different from the previous three are the multidisciplinary approach, automation, and flexibility. In combination with these modernized facets Industrial Revolution 4.0 seems to be the exact weapon to have in the repertoire of the medical practitioners across the world, keeping in mind the days to come after the ongoing pandemic [8–11].

### 1.1.2 Key elements that make Industry 4.0 extremely relevant in medical sciences:

Optimize the efficacy and productivity [12, 13]

Interpret the patient data of a huge proportion with minimal cost [14, 15]

Analyze the data with maximum accuracy [16]

Deliver the best quality [6, 7]

Improve both material and tool management [17, 18]

Maintain the clinical data in sophisticated computer-aided design (CAD) file instead of keeping in inventories [19, 20]

Produce customized implants in line with the patient requisite by developing smart AI-based sensors [21, 22]

Implement proper aids to sharpen the surgical interventions [23, 24]

With the advent of digital technology and sensor machineries the new diseases are autonomously detected with ease [25, 26]

Organize the centralized information database in the hospitals and other healthcare avenues [27–29]

Depict the medical data, relate with the relevant patient information [30, 31]

# 2 AI-based research/publication profiling in the recent past in line with the medical sciences: a brief outlook

AI-aided research and subsequent publications in relevance to the various wings of medical science have been one of the emerging fields since the last few years. Global publication databases such as *Scopus* and *Web of Science* already indexed round about 70 scholarly articles from 2013 to 2020 on the mentioned agenda. It is worth mentioning that in the context of ongoing global crisis the AI-based research in line with medical aspect becomes more significant as it portrays to substantial logic into that.

It began with the publication of the solitary article on the mentioned subject in 2013, with no further addition to that in 2014 and 2015. However, the number of original research significantly increased afterward; in the year 2016 there were six articles followed by another fourteen in 2017 in *Scopus* and *Web of Science*. It looked more promising as in 2018 the database showed another 34 research papers on the concerned area. By the year 2020 the trend continued to be in the right direction, with significant contributions coming from the AI-based research in tandem with the contemporary medical issues (Table 1).

**Table 1:** Frequency of AI-based research publications in line with medical practice in global databases from 2013 to 2020.

| Serial No. | Contributing field(s) | Frequency of publications (%) |
|---|---|---|
| 1. | Engineering sciences | 25 |
| 2. | Decision science | 11 |
| 3. | Mathematics | 11 |
| 4. | Business management and accounting | 06 |
| 5. | Material science | 04 |
| 6. | Biochemistry, genetics, and molecular biology | 02 |
| 7. | Miscellaneous (pharmacology, toxicology and pharmaceutics, physics and astronomy, psychology, and social sciences) | 10 |

**Source:** Retrieved from Scopus and Web of Science, as **of January 2020**

## 2.1 Web 2.0 and today's medical world: the making of a new era

The concept of Web 2.0 is not entirely convincing. It has created significant dilemma among the web experts across the globe. However, the concept has still been a pinnacle as far as the exchange of medical ideas are concerned [32]. According to the medical librarians Web 2.0 is the standout across the wings of the advanced medical world [33]. The implications of Web 2.0 becomes prominent in the context of the ongoing global pandemic where the work flow needs a continuous refinement and creative thinking in order to cope up with the norms of the new normal. Indeed, the medical research paid dividends over the last decades or so with the advent of Web 2.0, especially, during the transformation from Web 1.0 to Web 2.0 [34, 35]. Since the introduction of "digital narratives" the growth of Web 2.0 has been exponential in the field of biological sciences [36–44] (Table 2).

**Table 2:** Applications of Web 2.0 in medical fraternity.

| Application | Designated web portal | Significance |
|---|---|---|
| Bloglines | Bloglines | RSS reader |
| Citizendium | Citizendium | www.citizendium.org |
| Connotea | Connotea | Online reference organizer |
| Del.icio.us | http://del.icio.us | Website tagging |
| Flickr | www.flickr.com | Photo sharing |
| Ganfyd | www.ganfyd.org | Medical wiki |
| Google blogsearch | http://blogsearch.google.ca | Blog searches |
| Google health | www.google.com/coop/topics/Health | Create your own search tool |
| MedWorm | http://medworm.com | RSS aggregator |
| SlideShare | http://slideshare.net | Slide sharing |
| Wikipedia | http://en.wikipedia.org/wiki | All-purpose wiki |
| YouTube | www.youtube.com | Video snippets |

Web 2.0 took a rearguard action over the last few decades in refining the healthy lifestyle practice among the civilians. Moreover, it has been proactive in disease prevention, in diagnosis, and in treatment of several viral diseases [45]. It is beyond mention that application of Web 2.0 in the medical community is critical than ever before during the onslaught of the ongoing global pandemic and in the days to come. Since the outbreak of the COVID-19 pandemic the overall mindset of people has been severely bogged down due to an array of unavoidable circumstances. Innovative social media alongside Web 2.0 could be the silver lining under such tough times. With the user-friendly social aid people across all ages feel a state of comfort while sharing the information to the close ones. At the same point, they appear more expressive and involved with the proceedings by the influence of the digital platforms [46]. With due respect, conventional medical professionals, at times, miss to convey adequate information leading to the development of plenty of suspicious thinking among the patients. Modernized digital applications deliver the goods in reducing such paradox by providing detailed information on the disease concerned. Henceforth, the individuals have the option to go for crosschecking, verification, and even challenging the traditional prescriptions [46, 47].

In spite of the paradigm shift in the medical and healthcare sector with the emergence of digital innovations, still there have been both advantages and drawbacks existing in close proximities. Popular social platforms could be a double-edged sword in respect of the way they have to be used. In view the ongoing global crisis, therefore, it is even more vital to have enough knowledge and expertise about the disease taken into account before it is to be correlated with the Web 2.0. Both healthcare professionals and patients should take the responsibility to make the best out of this digital technology and minimize the margin of harmful effects associated with it.

# 3 Artificial intelligence-aided research in respective fields of medical practice

There has been a sizable proportion of AI-based literature existing in the modern-day healthcare practice; however, three major diseases have accounted for the lion's share of them: cancer, neurological disorders, and cardiovascular malfunctions. Another extremely critical aspect around which AI-based research has been going on of late are the different facets of genetic disorders. Indeed, AI-based research might be the future road map for the genetic research with the improvised concept and technologies (Figure 1).

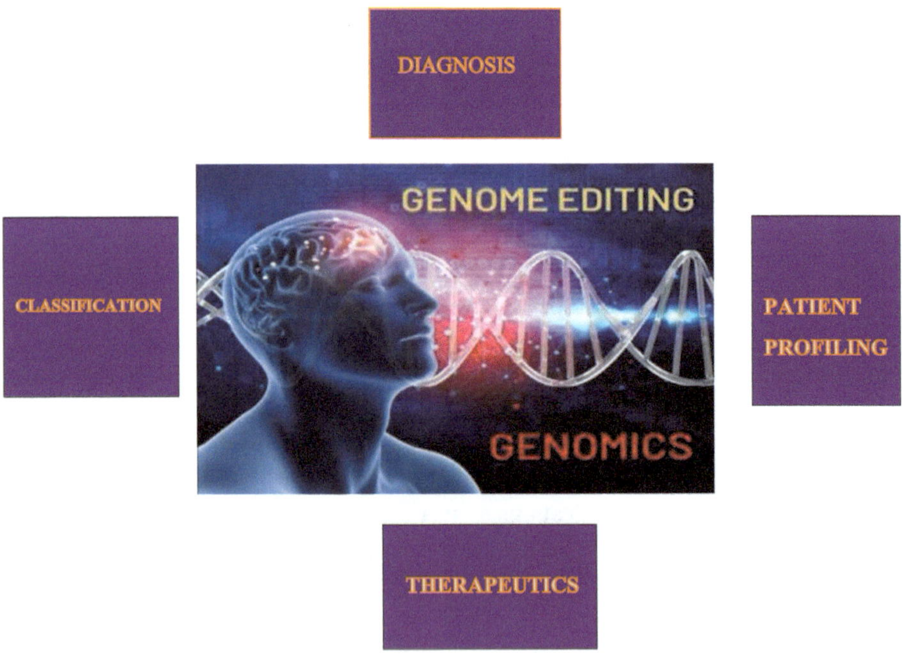


**Figure 1:** AI-mediated algorithms in the principal setups of genetic disease research.

## 3.1 Genetic disease: the fundamentals

The World Health Organization (WHO) categorized the genetic disorders as orphan diseases or rare diseases (RDs). RDs affect a minute portion of the population; however, with the overall prevalence of more than 7,000 varieties across the globe, genetic diseases are now constantly becoming the cause for concern [48]. Genetic disorders are primarily associated with the genomic aberrations (more than 80%), even though environmental predispositions including regular exposure to hazardous materials such as cosmic rays escalate the progression of such diseases [48, 49]. At the same point, intake of certain chemicals that could pass the transplacental barrier during the gestation period as well as infections is a leading factor [49]. For the last few decades the prevalence of RDs has been ever progressing. According to the recent data, around 350 million people across the globe have been estimated to be affected by the different varieties of such disorders [50, 51].

## 3.2 An insight on the limitations of genetic/rare disease research

Even though the genetic diseases are coming thick and fast in the global population, still it remains underrepresented as far as the overall output of the research is concerned. Following are the major attributes:

- Lack of appropriate identification of disease frequency in line with the improper notion of awareness, inadequate diagnostic protocols, and undefined clinical guidelines for the respective rare or genetic diseases [52, 53].
- Limited resources/funding to explore the hidden facets of such diseases [53].
- Relatively restricted or circumscribed geographical periphery regarding the availability of patients and, therefore, lack of volunteers have been always the hindrance to perform the clinical trials. Moreover, inappropriate coding system does contribute to the recruitment process of the patients [53, 54].
- The crucial factors adversely affecting the genetic disease research are the lack of collective knowledge, scarcity of data on the pathophysiology of the concerned disease respectively. Unavailability of disease specific bio-markers alongside underpowered clinical trials are the other obstacles [53].
- Lack of desired therapeutics, inadequate commercialization of available remedies and under per marketability are worth mentioning among the other contributing factors [54] (Table 3).

**Table 3:** Policies implemented by the respective countries to reduce the rare disease (RD).

| Strategic planning | Current status | Country(s) |
|---|---|---|
| Policy/plans | Decision to materialize specific planning | Indonesia, Singapore, Vietnam |
| | Very few meetings are conducted with the respective stakeholders | Thailand, Malaysia |
| | Policies in relevance to curb down the RDs submitted to the concerned ministry | Philippines, Australia and Canada |
| | Robust and well-formulated government policy existing in the rank of national setup | South Korea |
| | Strategies implemented completely nationwide | France, Taiwan |

Source: Retrieved from the web portal of World Health Organization (WHO) for country-wise statistics of RDs, as of August 2019

## 3.3 Perspective of genetic disease research from the recent past

In spite of the above-mentioned challenges, significant improvement has been achieved in the field of RD research with the advent of next generation sequencing and omics techniques. Both the technologies have contributed to the diagnosis and interpretation of the molecular mechanisms underneath [55, 56]. Furthermore, both of them have represented a handful of information about the existing RDs in order to analyze for the future proceedings [56]. The "age of big data" comprises a huge amount of data in relevance to the various wings of genetic/rare diseases; however, robust management skills as well as proper ethical guidelines are the paramount to achieve the goal.

### 3.3.1 Influence of AI on the genetic disease research of late

AI in tandem with machine learning (ML) has opened up a new window of opportunity to dissect and interpret the data by means of various subsets of algorithm. For the last decade several areas of modern-day medical practice have been benefited by the implementation of deep learning algorithms and the area of RD research is not an exception either. AI-based techniques were first introduced in medical research by the US Food and Drug Administration to systemize the clarity among the

food industries [57]. Machine-based algorithms are designed to analyze the complex data in the following manners [58]:
- By imposing supervised learning
- By imposing independent procedure
- By means of reinforcement

In the first of the above-mentioned categories the assigned algorithm determines the outcome by comparing the data that is put in with respect to the adjacent reference data. Henceforth, the result is obtained by the interpretation through the correlation of the respective data sets. Classification among various data types and regression analysis are the typical implementation areas of supervised learning [59, 60]. Supervised learning is routinely used these days in medical set up. Use of hematocrit to assess iron deficiency anemia is the prime example of that. On the other hand, in the second category that is independent of supervision, here information is put in without setting the predetermined goals or target. The designated algorithm performs autonomously to turn object data sets into cluster data [61, 62]. This technique is utilized in the evaluation of Alzheimer's disease and predicting the frequency of inheritable migraines. In reinforcement approach algorithms are direct in line with the precise action to be delivered to get the optimum outcome [63, 64]. Prediction of certain kind of epilepsy is the major outcome with respect to this advanced method.

# 4 Artificial intelligence (AI)-guided algorithms in various aspects of genetic disease research: an in-depth scenario

Of late, the application of AI-mediated algorithms has propelled the biomedicine to a new level. Various attributes of AI algorithms have been implied with the sequencing of disease pathogens, molecular modelling and drug designing leading to appropriate interpretation [65]. Following are the chief areas of interest with respect to the genetic/rare disease research where AI algorithms have been materialized (Figure 1):

A. Diagnosis and prognosis
B. Classification, identification/character profiling
C. Disease therapeutics
D. Patient profiling and health informatics

## 4.1 AI algorithms in the diagnostics of RDs

The key element in the diagnosis of RDs followed by the disease treatment and prophylaxis is the assessment of precise genes and mutations associated with it [66]. Whole-exome sequencing (WES) and whole genome sequencing (WGS) are the primary in vitro methods to deal with it in most of the advanced genetic research laboratories [67, 68]. The introduction of deep learning along with the various algorithms has made the procedures more user-friendly and cost-effective [67] (Figure 2).

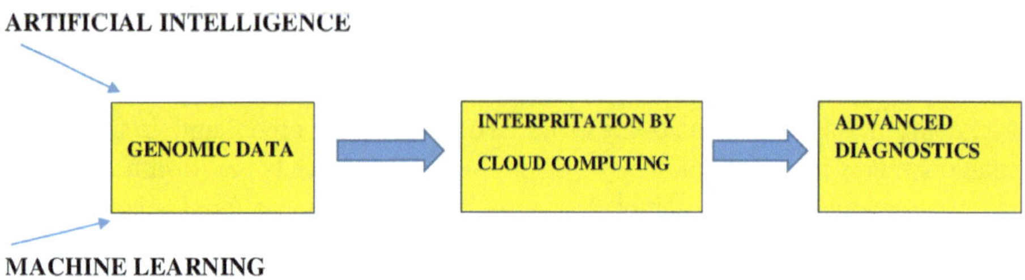


**Figure 2:** Schematic flow of AI-aided diagnostics for rare disease.

### 4.1.1 Identification of single nucleotide variants (SNVs)

Single nucleotide polymorphisms (SNPs) are the chief reasons behind most of the RDs. Therefore, assessment and subsequent analysis of such polymorphisms producing pathogenic aberration are pivotal for optimum usage of nuclear genome sequencing, especially, in precision medicine [67–69]. However, in the recent past, a handful of AI-generated techniques have made the things easier. SNVs are the major factors because of the abilities to change amino acid sequences leading to abnormal gene products. Relevant approaches like variant effect scoring tool (VEST) combined with random forest (RF) have been utilized to precisely assess missense mutations responsible for the alteration of protein activities [67]. Miller and Freeman Sheldon syndrome is the typical example where this method is implied to specifically fix the genes contributing to the abnormality with respect to the Mendelian genes. VEST is extremely secure as it performs without controlling factors. On the other hand, innovations such as CliniPred act using multiple of ML algorithms tagged with medical data to deliver high performance more efficiently. Again, SNV prediction tools are either structure- or sequence-based or the combination of both. This has been proven by the application of Weka that is developed especially to fix the RD data set from the overall protein database [70]. In the same regard, precisely generated supervised learning approaches have been also aligned to depict the degree of effects produced by the deleterious mutants in line with mevalonate kinase protein (ORPHA: 309025) [71]. Specific algorithms have been used to analyze the number of SNPs in common variable immunodeficiency (CVID) with significant success [72].

### 4.1.2 Identification of polygenic mutants

Introns are accountable for around 90% of the human RDs. This is suggestive of the fact that noncoding mutants are the major attributes of the RDs [73]. To evaluate such conditions a combination of splice AI and a deep residual neural network (NN) has been aligned to determine the deleterious cryptic splice mutants in the MYBPC3 noncoding sequence in association with hypertrophic cardiomyopathy (ORPHA:217569) [73]. This novel method has been an eye-opener in the diagnosis and interpretation in response to the cryptic splice mutants. RDs are complex as does the human genome itself. Bearing in mind that genetic diseases are predominantly polygenic in nature, the variant combinations pathogenicity predictor (VarCoPP) is formulated. It is an exclusively designed ML method for calculating the number of pathogenic gene pairs. VarCoPP is by far an accurate tool to have in the clinical setup; it offers 95% to 99% specific accuracy as far as the pathogenic scores are concerned [74].

### 4.1.3 Assessment of abnormal copy number variants

Human copy number variants (CNVs) are one of the productive markers in the diagnosis of RDs. Despite the presence of several readily available data bases, DNorm (specially designated AI-mediated data processing tool, governed by ML methods) has been developed. It generates named entities recognition that, in turn, produces CNV digest consisting of 1,582 CNVs and 2425 diseases respectively. This approach has been employed successfully in Di George Syndrome (ORPHA: 567) and has resulted in prudent outcomes [75].

### 4.1.4 Diagnosis through phenotype-governed approach

One of the classical improvisations among the ranks of phenotype-aided tools is the heterogeneous association network for RDs. This is a collective package of data-oriented algorithm-graphical convolution-mediated plotting in combination with disease-related genes, alleles, and phenotypes. This system renders the exact binary correlation among the existing terminologies. It has already delivered better performance in comparison to the conventional Orphamizers [76]. On the contrary, Phen–Gen is generated so as to inculcate data sequence from patients in regard to the symptoms obtained from recognized resources. It has provided remarkably authentic data for exons as well as introns with high accuracy. Furthermore, it surpasses the accuracy level by a significant 52% compared to the traditional protocols [77].

### 4.1.5 AI-based imaging in diagnosis of RDs

Morphological dysfunction is often characterized with RDs. AI-guided imaging has been one of the meaningful tools in assessing the phenotype data leading to the correct diagnosis [78, 79]. One of the convenient methods among the ranks is DeepGestalt, which uses the ML system in tandem with Face2Gene (a phenotype-driven framework) to recognize genetic disorders with moderate to high specificity [80]. Cornelia de Lange syndrome (ORPHA: 199), Emanuel syndrome (ORPHA: 96170), and Pallister–Killian syndrome (ORPHA: 884) are the examples where DeepGestalt has been successfully employed [81–83]. Similarly, AI algorithms along with clinical imaging techniques are designed to achieve significant images for acromegaly (ORPHA: 963) and Pick disease (ORPHA: 282) [84, 85].

### 4.1.6 Other associated aids

With the introduction of advanced methods the diagnosis of genetic disease is becoming easier. However, disorders like hemorrhagic telangiectasia (HHT, ORPHA: 774) still exhibit high cost for the diagnosis. Of late, neural networking has been tagged with infrared spectroscopy (IR) to minimize the testing cost for HHT [86]. Similarly, a deep learning predicting algorithm has been applied in patients with synovial sarcoma (SS, ORPHA: 3273). It obtained greater outcome in comparison to the conventional methods of diagnosis [86, 87]. Overall, various innovative ML techniques have added a new window to understand and depict the RDs with ease [87].

## 5 Discussion and conclusion

Unlike most of the conventional diseases ranging from the mild cough to the deadliest ones, genetic diseases are unique with respect to the trajectory, pathophysiology, and, of course, diagnosis. Until the advent of AI and ML algorithms medical practice had to depend on the traditional methods of gene sequencing and genetic counselling as the major diagnostic attributes. However, the introduction of AI-mediated procedures combined with patient phenotype-driven data has significantly excelled the optimum outcome toward RD diagnosis [88]. To illustrate more, AI-based medical diagnostics not only came with accurate diagnosis of RDs but also inferred precise treatment. More importantly, this new approach remarkably opened up the traits those were not diagnosed before [89, 90]. Besides this, it is quite understandable that genomic data cannot simply be the way out to deal with genetic diseases. To add to it, there is not enough data existing in terms of most of the RDs across the globe. Therefore, it is imperative to include AI and ML algorithms in modern genetic disease research to find out the hidden mechanisms associated with RDs leading to the development of novel therapeutics. All in all, the field of biomedicine is a huge one and there are several things to be remembered while introducing a new and computer-aided technology into practice. Henceforth, all the relevant issues including the social, ethical, as well as legal attenuations should be taken with full care and consideration in order to make AI-based medical practice a powerful weapon.

## References

[1] Murdoch, T.B. & Detsky, A.S. (2013). The inevitable application of big data to health care. JAMA, 309, 1351–1352.

[2] Kolker, E., Özdemir, V. & Kolker, E. (2016). How healthcare can refocus on its super-customers (Patients, n =1) and customers (Doctors and Nurses) by leveraging lessons from Amazon, Uber, and Watson. OMICS, 20, 329–333.

[3] Dilsizian, S.E. & Siegel, E.L. (2014). Artificial intelligence in medicine and cardiac imaging: harnessing big data and advanced computing to provide personalized medical diagnosis and treatment. Current Cardiology Reports, 16, 441.

[4] Patel, V.L., Shortliffe, E.H., Stefanelli, M. et al. (2009). The coming of age of artificial intelligence medicine. Artificial Intelligence in Medicine, 46, 5–17.

[5] Jha, S. & Topol, E.J. (2016). Adapting to Artificial Intelligence: radiologists and pathologists as information specialists. JAMA, 316, 2353–2354.

[6] Aceto, G., Persico, V. & Pescape, A. (2018). The role of Information and communication technologies in healthcare: taxonomies, perspectives, and challenges. Journal of Network and Computer Applications, https://doi.org/10.1016/j.jnca.2018.02.008.

[7] Pereira, T., Barreto, L. & Amaral, A. (2017). Network and information security challenges within Industry 4.0 paradigm. Procedia Manufacturing, 13, 1253–1260.

[8] Luthra, S. & Mangla, S.K. (2018). Evaluating challenges to Industry 4.0 initiatives for supply chain sustainability in emerging economies. Process Saf Environ Protect, https://doi.org/10.1016/j.psep.2018.04.018.

[9] Quezada, L.E., Da Costa, S.E.G. & Tan, K.H. (2017). Operational excellence towards sustainable development goals through Industry 4.0. International Journal of Production Economics, 190, 1–2.

[10] Haleem, A., Javaid, M. & Vaishya, R. (2019). Industry 4.0 and its applications in orthopedics. Journal of Clinical Orthopaedics and Trauma, 10(3), 615–616.

[11] Manogaran, G., Thota, C., Lopez, D. & Sundarasekar, R. (2017). Big Data Security Intelligence for Healthcare Industry 4.0. Cybersecurity for Industry 4.0., 715, 103–126.

[12] Wollschlaeger, M., Sauter, T. & Jasperneite, J. (2017). The future of industrial communication: automation networks in the era of the internet of things and industry 4.0. IEEE Industrial Electronics Magazine, 11(1), 17–27.
[13] Basl, J. (2017). Pilot study of readiness of Czech companies to implement the principles of Industry 4.0. Management and Production Engineering Review, 8(2), 3–8.
[14] Bahga, A. & Madisetti, V.K. (2013). A cloud-based approach for interoperable electronic health records (EHR's). IEEE Journal of Biomedical and Health Informatics, 17(5), 894–906.
[15] Wang, Z.G., Zhang, L. & Wen-Jun Zhao, W.J. (2017, Oct). Progress in digital medicine. Chinese Journal of Traumatology, 20(5), 297–298.
[16] Sommer, L. (2015). Industrial revolution-industry 4.0: are German manufacturing SMEs the first victims of this revolution? Journal of Industrial Engineering and Management, 8(5), 1512–1532.
[17] Lobo, F.A. (2017). Industry 4.0: Redefining Manufacturing of Future Medical Devices. ECN, https://www.ecnmag.com/article/2017/10/industry-40-redefiningmanufacturing-future-medical-devices.
[18] Ramanayake, R.P.J.C., Ranasinghe, S. & Lakmini, S. (2014). Management of emergencies in general practice: role of general practitioners. International Journal of Family Medicine and Primary Care, 3(4), 305–308.
[19] Wang, S., Wan, J., Zhang, D., Li, D. & Zhang, C. (2016). Towards smart factory for Industry 4.0: a self-organised multi-agent system with big data-based feedback and coordination. Computer Networks, 101, 158–168.
[20] Kilic, T. (2016). Digital hospital; an example of best practice. Journal of Health Services Research & Policy, 1(1), 52–58.
[21] Jee, K. & Kim, G.H. (2013). Potentiality of big data in the medical sector: focus on how to reshape the healthcare system. Healthcare Informatics Research, 19(2), 79–85.
[22] Liu, Y.K. & X., X. (2017). Industry4.0 and cloud manufacturing: a comparative analysis. Journal of Manufacturing Science & Engineering, 139(3), 1–8.
[23] Zhong, T.Y., Xu, X., Klotz, E. & Newman, S.T. (2017). Intelligent manufacturing in the context of industry 4.0: a review. Engineering, 3, 616–630.
[24] Javaid, M. & Haleem, A. (2018). Additive manufacturing applications in medical cases: a literature-based review. Alexandria Journal of Medicine, 54(4), 411–422.
[25] Kamble, S.S., Gunasekaran, A. & Gawankar, S.A. (2018). Sustainable industry 4.0 framework: a systematic literature review identifying the current trends and future perspectives. Process Safety and Environmental Protection, 117, 408–425.
[26] Mishra, S. (2017). Hologram the future of medicine e from Star Wars to clinical imaging. Indian Heart Journal, 69(4), 566–567.
[27] Brettel, M., Friederichsen, N., Keller, M. & Rosenberg, M. (2014). How virtualisation, decentralization and network building change the manufacturing landscape: an industry 4.0 perspective. Scientific Journal of Mechanical and Industrial Engineering, 8(1), 37–44.
[28] Theorin, A., Bengtsson, K., Provost, J. et al. (2017). An event-driven manufacturing information system architecture for Industry 4.0. International Journal of Production Research, 55(5), 1297–1311.
[29] Bahrin, M.A.K., Othman, M.F., Nor, N.H. & Azli, M.F.T. (2016). Industry 4.0: a review on industrial automation and robotic. Jurnal Teknologi, 78, 137–143.
[30] Javaid, M. & Haleem, A. (2018). 4D printing applications in medical field: a brief review. Clinical Epidemiology and Global Health, https://doi.org/10.1016/j.cegh.2018.09.007.
[31] Kotevski, A., Koceska, N. & Koceski, S. (2016) E-health monitoring system. In: International Conference on Applied Internet and Information Technologies. https://doi.org/10.20544/AIIT2016.32.

[32] O'Reilly, T. (2005). What is Web2.0? Design patterns and business models for the next generation of software. O'Reilly Media, www.oreillynet.com/pub/a/oreilly/tim/news/2005/09/30/what-is-web-20.html.
[33] Barsky, E. (2006). Introducing Web 2.0: weblogs and podcasting for health librarians. Journal of the Canadian Health Libraries Association, 27, 33–34.
[34] Shneiderman, B. (2008, Mar 7). Computer science. Science 2.0. Science, 319(5868), 1349–1350.
[35] Whitmire, A. (2013). Thoughts on "eResearch": a Scientist's Perspective. JESLIB, 2(2), 68–72.
[36] Giustini, D. (2006, Dec 23). How Web 2.0 is changing medicine. BMJ, 333(7582), 1283–1284.
[37] Boudry, C. (2012). ["Biology/medicine 2.0": an overview]. Medecine Sciences : M/S, 28(6–7), 653–658.
[38] Van De Belt, T.H., Engelen, L.J., Berben, S.A. & Schoonhoven, L. (2010). Definition of Health 2.0 and Medicine 2.0: a systematic review. Journal of Medical Internet Research, 12(2), e18.
[39] Eysenbach, G. (2008). Medicine 2.0: social networking, collaboration, participation, apomediation, and openness. Journal of Medical Internet Research, 10(3), e22.
[40] Oh, H., Rizo, C., Enkin, M. & Jadad, A. (2005). What is eHealth? a systematic review of published definitions. World Hospitals and Health Services Journal, 41(1), 32–40.
[41] Gardois, P., Colombi, N., Grillo, G. & Villanacci, M.C. (2012). Implementation of Web 2.0 services in academic, medical and research libraries: a scoping review. Health Information and Libraries Journal, 29(2), 90–109.
[42] Collins, E. & Hide, B. Use and relevance of web 2.0 resources for researchers. In: Publishing in the networked world: Transforming the Nature of Communication. Turid Hedlund and Yasar Tonta; 2010 Presented at: 14th International Conference on Electronic Publishing; 16–18 June 2010; Helsinki, Finland p. 271–289.
[43] Procter, R., Williams, R., Stewart, J., Poschen, M., Snee, H. & Voss, A. (2010). Adoption and use of Web 2.0 in scholarly communications. Philosophical Transactions of The Royal Society A Mathematical Physical and Engineering Sciences, 368(1926), 4039–4056.
[44] Cdcfoundation.org. What is Public Health? CDC Foundation. [Accessed 6 Mar. 2018]. [online] Available at: https://www.cdcfoundation.org/what-public-health
[45] Lin, W.-Y., Zhang, X., Song, H. & Omori, K. (2016). Health information seeking in the Web 2.0 age: Trust in social media, uncertainty reduction, and self-disclosure. Computers in Human Behaviour, 56, 289–294.
[46] Moorhead, S.A., Hazlett, D.E., Harrison, L., Carroll, J.K., Irwin, A. & Hoving, C. (2013, April). A new dimension of health care: Systematic review of the uses, benefits, and limitations of social media for health communication. Journal of Medical Internet Research.
[47] Sernadela, P., González-Castro, L., Carta, C., Van Der Horst, E., Lopes, P., Kaliyaperumal, R., Thompson, M., Thompson, R., Queralt-Rosinach, N., Lopez, E. et al. (2017). Linked registries: connecting rare diseases patient registries through a semantic web layer. BioMed Research International, 2017, 1–13.
[48] Ekins, S. (2017). Industrializing rare disease therapy discovery and development. Nature Biotechnology, 35, 117–118.
[49] About Rare Diseases. Available online: https://www.eurordis.org/about-rare-diseases. (accessed on 18 September 2019).
[50] Ronicke, S., Hirsch, M.C., Turk, E., Larionov, K., Tientcheu, D. & Wagner, A.D. (2019). Can a decision support system accelerate rare disease diagnosis? Evaluating the potential impact of Ada DX in a retrospective study. Orphanet Journal of Rare Diseases, 14, 69.
[51] Nestler-Parr, S., Korchagina, D., Toumi, M., Pashos, C.L., Blanchette, C., Molsen, E., Morel, T., Simoens, S., Kaló, Z., Gatermann, R. et al. (2018). Challenges in research and health

technology assessment of rare disease technologies: report of the ispor rare disease special interest group. Value Health, 21, 493–500.
[52] Stoller, J.K. (2018). The Challenge of Rare Diseases. Chest, 153, 1309–1314.
[53] Ahmed, M.A., Okour, M., Brundage, R. & Kartha, R.V. (2019). Orphan drug development: The increasing role of clinical pharmacology. Journal of Pharmacokinetics and Pharmacodynamics, 46, 395–409.
[54] Kaufmann, P., Pariser, A.R. & Austin, C. (2018). From scientific discovery to treatments for rare diseases – The view from the National Center for Advancing Translational Sciences – Oce of Rare Diseases Research. Orphanet Journal of Rare Diseases, 13, 196.
[55] Schee Genannt Halfmann, S., Mählmann, L., Leyens, L., Reumann, M. & Brand, A. (2017). Personalized Medicine: What's in it for Rare Diseases? In: Posada De La Paz, M., Taruscio, D. & Groft, S.C. (Eds.), Rare Diseases Epidemiology: Update and Overview, Cham, Switzerland, Springer International Publishing, Vol. 1031, 387–404, ISBN 978-3-319-67142-0.
[56] Toh, T.S., Dondelinger, F. & Wang, D. (2019). Looking beyond the hype: Applied AI and machine learning in translational medicine. EBioMedicine, 47, 607–615.
[57] Bishop, C.M. (2006). Pattern Recognition and Machine Learning, Information Science and Statistics, New York, NY, Springer, ISBN 978-0-387-31073-2.
[58] Pasa, F., Golkov, V., Pfei_er, F., Cremers, D. & Pfeiffer, D. (2019). Efficient deep network architectures for fast chest X-Ray tuberculosis screening and visualization. Scientific Reports, 9, 6268.
[59] Portnoi, T., Yala, A., Schuster, T., Barzilay, R., Dontchos, B., Lamb, L. & Lehman, C. (2019). Deep Learning Model to Assess Cancer Risk on the Basis of a Breast MR Image Alone. The American Journal of Roentgenology, 213, 227–233.
[60] Alashwal, H., El Halaby, M., Crouse, J.J., Abdalla, A. & Moustafa, A.A. (2019). The application of unsupervised clustering methods to Alzheimer's disease. Frontiers in Computational Neuroscience, 13, 31.
[61] Yang, M.-H., Yang, F.-Y. & Oyang, Y.-J. (2013). Application of density estimation algorithms in analyzing co-morbidities of migraine. Netw Model Anal Health Inform Bioinform, 2, 95–107.
[62] Lunceford, J.K., Davidian, M. & Tsiatis, A.A. (2002). Estimation of survival distributions of treatment policies in two-stage randomization designs in clinical trials. Biometrics, 58, 48–57.
[63] Panuccio, G., Guez, A., Vincent, R., Avoli, M. & Pineau, J. (2013). Adaptive control of epileptiform excitability in an in vitro model of limbic seizures. Experimental Neurology, 241, 179–183.
[64] Topol, E.J. (2019). High-performance medicine: The convergence of human and artificial intelligence. Nature Medicine, 25, 44–56.
[65] Carter, H., Douville, C., Stenson, P.D., Cooper, D.N. & Karchin, R. (2013). Identifying mendelian disease genes with the variant E ect scoring tool. BMC Genomics, 14, S3.
[66] Alirezaie, N., Kernohan, K.D., Hartley, T., Majewski, J. & Hocking, T.D. (2018). ClinPred: prediction tool to identify disease-relevant nonsynonymous single-nucleotide variants. American Journal of Human Genetics, 103, 474–483.
[67] Bosio, M., Drechsel, O., Rahman, R., Muyas, F., Rabionet, R., Bezdan, D., Domenech Salgado, L., Hor, H., Schott, J., Munell, F. et al. (2019). eDiVA – Classification and prioritization of pathogenic variants for clinical diagnostics. Human Mutation, 40, 865–878.
[68] Sundaram, L., Gao, H., Padigepati, S.R., McRae, J.F., Li, Y., Kosmicki, J.A., Fritzilas, N., Hakenberg, J., Dutta, A., Shon, J. et al. (2018). Predicting the clinical impact of human mutation with deep neural networks. Nature Genetics, 50, 1161–1170.
[69] Dehiya, V., Thomas, J. & Sael, L. (2018). Impact of structural prior knowledge in SNV prediction: Towards causal variant finding in rare disease. PLoS ONE, 13, e0204101.

[70] Browne, C. & Timson, D.J. (2015). In Silico Prediction of the E ects of Mutations in the Human Mevalonate Kinase Gene: Towards a Predictive Framework for Mevalonate Kinase Deficiency: E ects of Mutations in Human MVK. Annals of Human Genetics, 79, 451–459.
[71] Orange, J.S., Glessner, J.T., Resnick, E., Sullivan, K.E., Lucas, M., Ferry, B., Kim, C.E., Hou, C., Wang, F., Chiavacci, R. et al. (2011). Genome-wide association identifies diverse causes of common variable immunodeficiency. The Journal of Allergy and Clinical Immunology, 127, 1360–1367.e6.
[72] Jaganathan, K., Kyriazopoulou Panagiotopoulou, S., McRae, J.F., Darbandi, S.F., Knowles, D., Li, Y.I., Kosmicki, J.A., Arbelaez, J., Cui, W., Schwartz, G.B. et al. (2019). Predicting splicing from primary sequence with deep learning. Cell, 176, 535–548.e24.
[73] Papadimitriou, S., Gazzo, A., Versbraegen, N., Nachtegael, C., Aerts, J., Moreau, Y., Van Dooren, S., Nowé, A., Smits, G. & Lenaerts, T. (2019). Predicting disease-causing variant combinations. Proceedings of the National Academy of Sciences USA, 116, 11878–11887.
[74] Yang, X., Song, Z., Wu, C., Wang, W., Li, G., Zhang, W., Wu, L. & Lu, K. (2018). Constructing a database for the relations between CNV and human genetic diseases via systematic text mining. BMC Bioinformatics, 19, 528.
[75] Rao, A., Vg, S., Joseph, T., Kotte, S., Sivadasan, N. & Srinivasan, R. (2018). Phenotype-driven gene prioritization for rare diseases using graph convolution on heterogeneous networks. BMC Medical Genomics, 11, 57.
[76] Javed, A., Agrawal, S. & Ng, P.C. (2014). Phen-Gen: Combining phenotype and genotype to analyze rare disorders. Nat Methods, 11, 935–937.
[77] Smpokou, P., Lanpher, B.C. & Rosenbaum, K.N. (2015). Important considerations in the initial clinical evaluation of the dysmorphic neonate. Adv Neonatal Care, 15, 248–252.
[78] Smith, K., Piccinini, F., Balassa, T., Koos, K., Danka, T., Azizpour, H. & Horvath, P. (2018). Phenotypic image analysis software tools for exploring and understanding big image data from cell-based assays. Cell Systems, 6, 636–653.
[79] Gurovich, Y., Hanani, Y., Bar, O., Nadav, G., Fleischer, N., Gelbman, D., Basel-Salmon, L., Krawitz, P.M., Kamphausen, S.B., Zenker, M. et al. (2019). Identifying facial phenotypes of genetic disorders using deep learning. NatMed, 25, 60–64.
[80] Zarate, Y.A., Smith-Hicks, C.L., Greene, C., Abbott, M.-A., Siu, V.M., Calhoun, A.R.U.L., Pandya, A., Li, C., Sellars, E.A., Kaylor, J. et al. (2018). Natural history and genotype-phenotype correlations in 72 individuals with SATB2 -associated syndrome. American Journal of Medical Genetics, 176, 925–935.
[81] Basel-Vanagaite, L., Wolf, L., Orin, M., Larizza, L., Gervasini, C., Krantz, I.D. & Deard, M.A. (2016). Recognition of the Cornelia de Lange syndrome phenotype with facial dysmorphology novel analysis: Recognition of the CdLS Phenotype with FDNA. Clinical Genetics, 89, 557–563.
[82] Liehr, T., Acquarola, N., Pyle, K., St-Pierre, S., Rinholm, M., Bar, O., Wilhelm, K. & Schreyer, I. (2018). Next generation phenotyping in Emanuel and Pallister-Killian syndrome using computer-aided facial dysmorphology analysis of 2D photos: LIEHR et al. Clinical Genetics, 93, 378–381.
[83] Learned-Miller, E., Lu, Q., Paisley, A., Trainer, P., Blanz, V., Dedden, K. & Miller, R. (2006). Detecting Acromegaly: Screening for Disease with a Morphable Model. In: Larsen, R., Nielsen, M. & Sporring, J. (Eds.), Medical Image Computing and Computer-Assisted Intervention – MICCAI 2006, Heidelberg, Germany, Springer, Vol. 4191, 495–503, ISBN 978-3-540-44727-6.
[84] Spinelli, E.G., Mandelli, M.L., Miller, Z.A., Santos-Santos, M.A., Wilson, S.M., Agosta, F., Grinberg, L.T., Huang, E.J., Trojanowski, J.Q., Meyer, M. et al. (2017). Typical and atypical pathology in primary progressive aphasia variants: Pathology in PPA Variants. Annals of Neurology, 81, 430–443.

[85] Lux, A., Müller, R., Tulk, M., Olivieri, C., Zarrabeita, R., Salonikios, T. & Wirnitzer, B. (2013). HHT diagnosis by Mid-infrared spectroscopy and artificial neural network analysis. Orphanet Journal of Rare Diseases, 8, 94.
[86] Han, I., Kim, J.H., Park, H., Kim, H.-S. & Seo, S.W. (2018). Deep learning approach for survival prediction for patients with synovial sarcoma. Tumor Biology, 40, 101042831879926.
[87] Van Ijzendoorn, D.G.P., Szuhai, K., Briaire-de Bruijn, I.H., Kostine, M., Kuijjer, M.L. & Bovée, J.V.M.G. (2019). Machine learning analysis of gene expression data reveals novel diagnostic and prognostic biomarkers and identifies therapeutic targets for soft tissue sarcomas. PLoS Computational Biology, 15, e1006826.
[88] Wang, F. & Preininger, A. (2019). AI in health: state of the art, challenges, and future directions. Yearbook of Medical Informatics, 28, 016–026.
[89] Conejar, R.J. & Kim, H.-K. (2014). A Medical Decision Support System (DSS) for ubiquitous healthcare diagnosis system. International Journal of Software Engineering & Applications, 8, 8.
[90] Jurmeister, P., Schöler, A., Arnold, A., Klauschen, F., Lenze, D., Hummel, M., Schweizer, L., Bläker, H., Pfitzner, B.M., Mamlouk, S. et al. (2019). DNA methylation profiling reliably distinguishes pulmonary enteric adenocarcinoma from metastatic colorectal cancer. Modern Pathology, 32, 855–865.

Purvi Pujari, Priyeta Priyadarshini

# Knowledge Management: A Driver for Boosting Intellectual Capital

**Abstract:** The future belongs to the knowledge-based economies. A country's growth potential will be increasingly dependent on the knowledge potential and management of the knowledge processes by the same. As the globe gears toward a circular economy, the aspects of sustainability and innovation become very crucial. The need of the hour is to focus on intellectual capital (IC) as a key driver of the knowledge-based economy. In an economy, the focus should be on creating and developing knowledge intensive industries. The method is to prioritize the boosting of IC and enhance value creation in the economy. The correct emphasis on knowledge management (KM) in an economy can be leveraged through boosting IC. This will lead to a more sustainable and innovative nation moving toward a circular economy. For any organization to ensure sustainable competitive advantage it is imperative to weave KM in its human resource initiatives. Investing in KM is a long-term strategy. Technology plays a crucial role in this whole process. KM technologies are information technologies that can be used to facilitate KM. The aim of this chapter is to identify the value creation process while providing suggestions for strategic interventions for a fruitful relation between an organization's employees, suppliers, customers, and other stakeholders. This process needs interlinking of technology with the human resource processes. This chapter is divided into three main parts. The first part focuses on defining and explaining types of KM wherein the importance of KM systems is detailed. The second part deals with describing IC, its significance and basic components. The third and most important part deals with the relationship between KM and IC and suggests strategies for boosting IC. As the tertiary sector becomes more and more important and Technology 4.0 becomes the key driver of every growth activity, knowledge acquisition and KM will be the game changers.

**Keywords:** Knowledge Management, Intellectual Capital, Technology 4.0, Strategy

## 1 Introduction

The move from a traditional economy to an information-based economy requires a consistent focus on knowledge management (KM) in society. The organizations in any society have to emphasize value creation through an impetus to intellectual capital (IC). KM is a deliberate process of identifying, structuring, maintaining, and exchanging the expertise and experience of employees within an organization. It is an intricate process of value creation out of the knowledge potential of an organization. It

https://doi.org/10.1515/9783110722789-008

involves several elements like human resources practices, technology, culture, and organizational structures. There have been numerous attempts to define KM. KM is defined as "the process of creating, sharing, using and managing the knowledge and information of an organization [5]. It refers to a multidisciplinary approach to achieve organizational objectives by making the best use of knowledge." KM is the name of a concept in which an enterprise consciously and comprehensively gathers, organizes, shares, and analyzes its knowledge in terms of resources, documents, and people skills. KM draws from existing resources that your organization may already have in place-good information systems management, organizational change management, and human resources management practices.

IC has been defined as "Intellectual capital is the term given to the combined intangible assets which enable the company to function." IC is the intangible value of a business, covering the competencies of its people (human capital), the value relating to its relationships (relational capital), and everything that is left when the employees go home [2]. The focus of this chapter is to explore the role of KM in boosting IC.

KM is becoming the key function to enhance the sustainable competitive value of an organization. It refers to the learning and development of a company or its clients. It consists of a cycle of information formation, sharing, structuring, and auditing to optimize the efficacy of an organization.

## Knowledge management has the following knowledge-focused activities:

1. Generation of new knowledge or information
2. Acquiring important information from external sources
3. Utilizing available knowledge and information in intelligent decision-making
4. Implanting knowledge in product and services
5. Constituting knowledge and information in databases and respective software
6. Assisting in the growth of knowledge through culture and motives
7. Conveying information of the current organization into other parts of the organization
8. Impact of KM

The importance of KM is that it contributes maximum to the development of any organization and nation as a whole. Currently, the world is observing an information explosion. Knowledge allows the creation of abilities that determines the capacity to do things in an effective manner. Knowledge allows making decisions, predictions, and associations.

Knowledge is the main source of decision-making, planning, design, and evaluation. Knowledge is also important expertise which creates a difference to individuals, groups of employees, and organizations. Generating, imparting, and disseminating knowledge or information is critical and it should be handled with care.

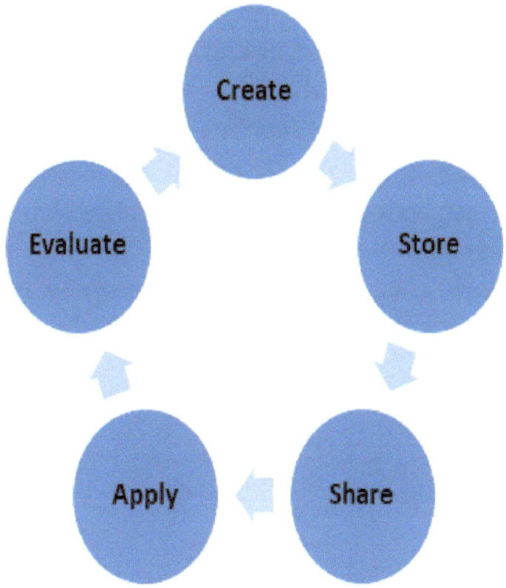


**Figure 1:** Process of knowledge management. (Prepared by the authors)

## There are seven knowledge levels in knowledge management:

1. Knowledge of customers
2. Knowledge of people/employee
3. Knowledge of products and services
4. Knowledge of processes
5. Organizational database
6. Knowledge of relationships to deal with customers, suppliers, and stakeholders
7. Knowledge about IC of the workforce

## Steps in KM process:

The KM process comprises following three important areas:

1) **Accumulation of knowledge** – There are various models which have been created for accumulation of knowledge. Every organization must ensure the cooperation of employees in mind while selecting the models. Many a times, employees are resistant to change and are not willing to cooperate. Knowledge accumulation is a gradual process and has to be done in stages. All resources within the organization must be thoroughly studied to ensure a comprehensive knowledge acquisition. The more important aspect for any organization is to prioritize its requirements and select the suitable technique.

2) **Storage of information** – Any KM system must have the proper selection of its devices and other technology parameters. The capturing of knowledge becomes a significant step where the pathways and systems in the organizational memories are defined. This is also known as institutionalization of the knowledge.
3) **Sharing of knowledge** – This is an area which specifically deals with distribution of knowledge. In this process, the interaction of employees among themselves and other stakeholders plays a very important part. The inherent culture of the organization is also very significant in ensuring the successful encompassing of the knowledge.

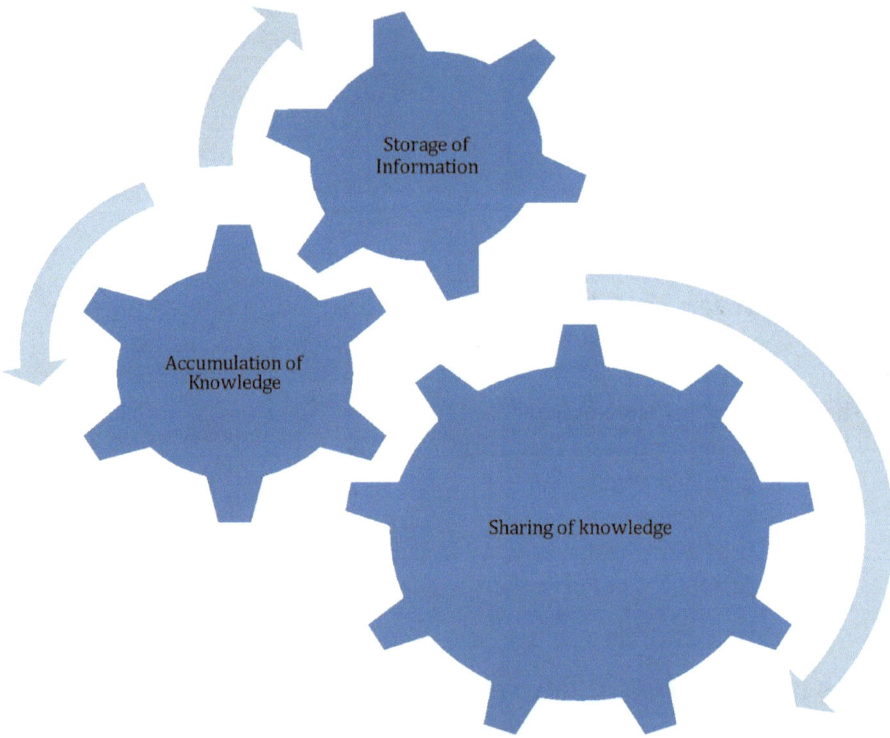


**Figure 2:** Steps in knowledge management process. (Prepared by the authors)

The main goal of KM is to initiate a learning culture and empower organizational learning. Sharing of knowledge is also encouraged among the employees. There are mainly three types of knowledge that need to be considered while we discuss KM:

### Types of knowledge:

1. **Tacit knowledge** – This knowledge is inborn and natural. It is not easily codified and sharing is difficult. Examples of such knowledge are creative thinking, analytical thinking. This knowledge is gained usually from personal experience and it is considered as difficult when it comes to sharing knowledge.
2. **Explicit knowledge** – It is easy to share, understand, teach, and articulate. Examples include solving a mathematical equation, how to download an application in android.
3. **Implicit knowledge** – It is nothing but the application of explicit knowledge. The skills which can be transferable from one work to another are an example of implicit knowledge.

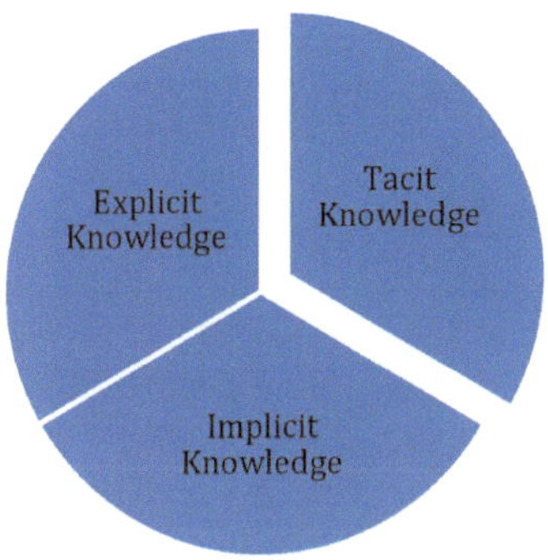


**Figure 3:** Types of knowledge. (Prepared by the authors)

## 2 Types of knowledge management

### There are basically four types of knowledge management, which are as follows:

a) **Internal KM** – It is a method of giving the staff admittance to information in order to help them to manage their responsibilities better or to improve overall information. This can incorporate different kinds of instruments and approaches – intranets, tool boxes, information bases of exploration or exercises learned, networks of training, information sharing occasions.

b) **Information dispersal** – It is the process of creating information and making the hierarchical information as generally available, or known to the outside world particularly to the stakeholders. For associations with a solid base either in research, or in practical experience, needs to disseminate the information which is broadly accessible, open, and utilized as could be expected under the circumstances.

c) **Information facilitation** – It is the process of associating the stakeholders with significant information and expertise. Here the job is to help interface advancement accomplices and partners with the information they need, regardless of whether it comes from internal or external sources.
d) **Building information limit** – It is the process of building the limit of organizational clients to create, get, offer, and use information viably. This may be the most testing yet additionally most principal approach to put information at the administration of improvement.

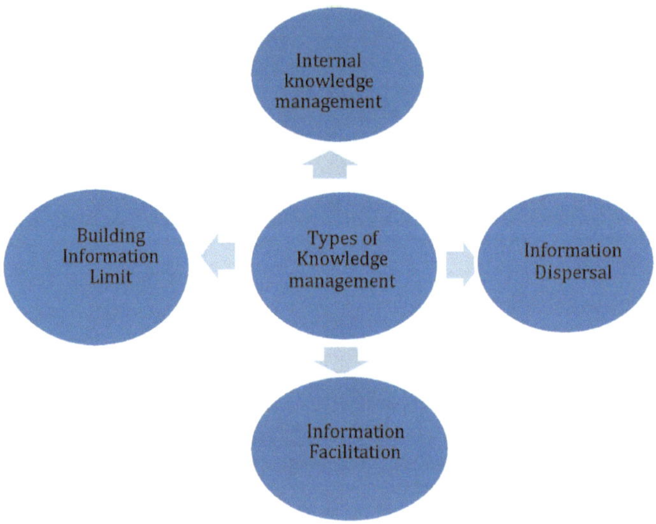


**Figure 4:** Types of knowledge management. (Prepared by the authors)

# 3 Components of intellectual capital

1. **Human capital** – The creative capacity, awareness, ability, and knowledge of employees is the human capital of any organization. Although people do not like to be denoted themselves as capital, so terms like talent and human resources are used alternatively.
2. **Relational capital** – It explains the relationship of an organization with investors, customers, employees, stakeholders, and communities. This includes both formal and informal contracts.
3. **Structural capital** – Company's capacity to invent and innovate, business processes and culture contribute to structural capital. These are intangible assets of any organization. Examples include data, media, intellectual property, systems and processes.

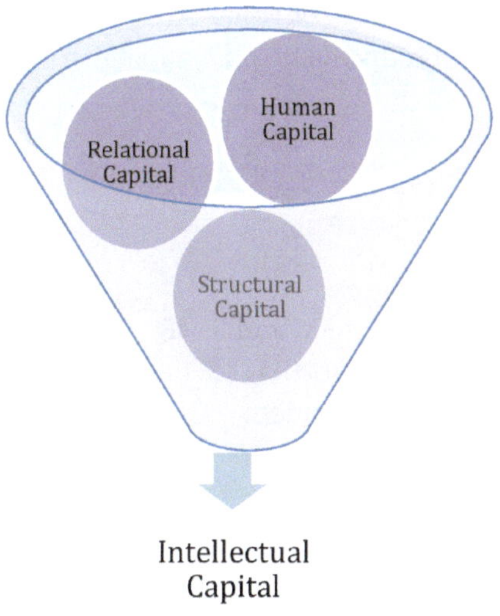


**Figure 5:** Components of intellectual capital. (Prepared by the authors)

The achievement of each organization depends increasingly more on human resources than just physical capital. At the point when we think about human resources as a mix of capability, aptitudes, capacities and insight of individuals, that implies as a factor, that gives exceptional, unique, and uncommon character to each business, at that point, we realize that specifically, representatives are the source that can learn, ready to make changes, development, and willingness to have innovative exertion, which directly can lead to long-term flourishing organization.

IC is a term that refers to the intangible assets of an organization, its employees, and their cumulative knowledge and awareness about a function/job. It includes its relationships with its investors, customers, which is termed as relationship capital, its employees, employee training and development, performance appraisal which is termed as human capital, and what is left when employees go home, that is, database, policies, culture, and mission is termed as structural capital. It is a conceptual term used to try to measure assets a company may have that are not visible on a balance sheet, such as the expertise and experience of its employees.

Although this is an important component of any organization, it is subjective and very hard to quantify accurately. However, successful organizations usually focus on uplift and enhance their IC by training their employees to a certain level and making sure they are motivated and encouraged to give their best in a congenial and conducive environment. Companies that fail to take these thoughts into account may find employees who are less motivated, poorly trained, thereby affecting profits and productivity of the organization.

Hence, the focus on IC is very crucial for the process of value creation of any organization [1]. This in turn impacts the overall productivity and competitive sustainability of the company in the long-term.

# 4 Literature review

Researchers across the world have studied the concept of KM and its significance in boosting IC. The following literature review Table 1 suggests that KM has been identified as an effective tool in enhancing the IC of any organization.

**Table 1:** Literature review.

| S. no. | Author | Take on KM & IC |
|---|---|---|
| 1. | Jan et al. [7] | IC statement to be used as a communication regarding knowledge-based strategy. |
| 2. | Garcia-Perez et al. [4] | The focus of this research has been on the calculation and reporting of IC. |
| 3. | Marr [9] | This book gives a multidimensional perspective on the concept of IC. |
| 4. | Yusliza et al. [13] | This study focuses on the green IC as a model to increase clean production in the long term. |
| 5. | Richard & James [12] | This is an extensive research work focusing on the literature pertaining to IC. It underlines the significance of IC for the progress of any economy. |
| 6. | Leif & Patrick [3] | This work is focused on the definition and explanation of fundamental concepts related to IC and KM. |
| 7. | Nick [11] | This research work explores the concept of growth in business and IC. The performance of a company and its relation with the IC has been studied in this study. |
| 8. | Mohan & Youndt [10] | The capabilities embedded in any organization have been extensively studied in this research along with the association with the IC. It also examines the innovation performance of a business as a result of enhancement in IC. |
| 9. | Antonio [1] | This study states that as the earlier concept of organizational capital changes and relevance of IC takes a front seat. The organization must give more focus to report components of IC and ensure its disclosure. |
| 10. | Hussinki et al. [6] | This research focuses on an organization's performance link with its management of IC and KM. |

The detailed review of the literature suggests that the pre-existing field of KM has taken a center stage with technology and human capital becoming the revenue generation sources. The disclosure of IC by the organizations is becoming crucial day-by-day [1]. There have been lots of studies suggesting the close link between IC and KM and the performance of an organization [6]. The progress of an organization is closely related to the IC assets it holds and also the way it manages them with leading technologies of KM [11]. To make this process effective, there is an imperative need to redefine the concepts related to IC and KM [8]. The connect between innovation and the business performance of the organization has been studied by many researchers. Sustainability is a new field that has been increasingly been explored by researchers all over the globe. There has been an increased emphasis on upgrading technological infrastructure.

# 5 Knowledge management: a driver for boosting for intellectual capital

The process of improving competitive sustainability lies strongly on the boost to IC using KM as the key driver. The focus is more and more on tacit knowledge. The working together of people, processes, and technology is a must to build a sustainable strategic business position. To ensure the process of value creation is continued in an organization, the use of technology and the existence of a clear vision is a must. The top management must foster a culture of knowledge sharing and encompassing the current technology to boost IC.

# 6 Strategies for boosting IC

Any organization which wants to excel in this competitive world has to leverage its strength in IC. There can be multiple ways to boost IC in an organization. Following are the strategies to enhance IC:

1. **Establishing a firm culture of knowledge enhancement:** For enhancing IC in any firm, there has to be a culture of giving priority to knowledge enhancement. Establishing such a culture will require the policy guidelines from the top management which underlines the value of knowledge sharing and a boost to related competencies and skills.
2. **Establish a system to integrate ethics into your organization's culture:** Social capital of any organization cannot be improved unless and until there is the due significance given to ethics in the process of accumulation of knowledge.

3. **Create the infrastructure of KM and retention:** Every organization must invest resources to establish and communicate the same to its stakeholders.
4. **Holistic development of human capital:** Any process of developing IC in an organization is incomplete and ineffective unless and until it focuses on the holistic development of its employees. The structure and process of IC have to encompass all aspects of the development of a human being.
5. **Connect with organizational strategic intent:** The KM infrastructure has to be up-to-date and consistent with the requirements of the organization. A strong connection has to be ensured with organizational long-term strategies and vision.

# 7 Conclusion

In the post-COVID-19 world, the relative significance of the IC of an organization has been proven once again. As the traditional sources of competitiveness of an organization take a back seat and relevance of IC has been reestablished. The firms must rethink and re-strategize to give a boost to their IC. The knowledge domains have to be clearly identified and should be given due importance. The companies have to identify and reinforce a well-defined strategy for boosting IC. This strategy should be designed keeping in mind the vision and ethical culture of the organization. In the coming days, the organization with a sound ethical IC will be able to withstand the turbulence in the corporate world. The future belongs to an organization with an efficient KM infrastructure. This will generate a sustainable competitive edge in the global business landscape. Intellectual and social capital will be the new currency in the days to come.

# Bibliography

[1] Salvi, A., Vitolla, F., Giakoumelou, A., Raimo, N. & Rubino, M. (2020). Intellectual capital disclosure in integrated reports: The effect on firm value. In: Technological Forecasting and Social Change, Elsevier, Vol. 160. doi: 10.1016/j.techfore.2020.120228.

[2] Edvinsson, L. & Malone, M.S. (1997). Intellectual Capital: Realizing Your Company's True Value by Finding Its Hidden Brainpower, New York, Harper Business.

[3] Durst, S. & Zieba, M. (2019). Mapping knowledge risks: towards a better understanding of knowledge management. Knowledge Management Research & Practice, 17(1), 1–13. https://doi.org/10.1080/14778238.2018.1538603.

[4] Garcia-Perez, A., Ghio, A., Occhipinti, Z. & Verona, R. (2020). Knowledge management and intellectual capital in knowledge-based organisations: a review and theoretical perspectives. Journal of Knowledge Management, 24(7), 1719–1754. https://doi.org/10.1108/JKM-12-2019-0703.

[5] Girard, J.P. & Girard, J.L. (2015). Defining knowledge management: Toward an applied compendium. Online Journal of Applied Knowledge Management, 3, 1. doi: 10.1016/j.sbspro.2018.03.022.
[6] Hussinki, H., Ritala, P., Vanhala, M. & Kianto, A. (2017). Intellectual capital, knowledge management practices and firm performance. Journal of Intellectual Capital, 18(4), 904–922. 2., https://doi.org/10.1108/JIC-11-2016-0116.
[7] Jan, M., Per, N.B. & Bernard, M. (2020). Reporting on intellectual capital: why, what and how? Emerald Group Publishing Limited, 8(1), 46–54.
[8] Leif, E. & Patrick, S. (1996). Developing a model for managing intellectual capital. European management Journal, 14(4), 356–364. https://doi.org/10.1016/0263-2373(96)00022-9.
[9] Marr, B. (Ed). (2005). Perspectives on Intellectual Capital. Multidisciplinary Insights into Management, Measurement and Reporting, Oxford, Elsevier.
[10] Mohan, S. & Youndt, M.A. (2005). The Influence of Intellectual Capital on the Types of Innovative Capabilities. Academy of Management Journal, 48(3), https://doi.org/10.5465/AMJ.2005.17407911.
[11] Nick, B. (1998). Intellectual capital: an exploratory study that develops measures and models. Management Decision, 36(2), 63–76. 6., https://doi.org/10.1108/00251749810204142.
[12] Richard, P. & James, G. (2000). Intellectual capital literature review: Measurement, reporting and management. Journal of Intellectual Capital, 1(2), 155–176. doi: 10.1108/14691930010348731.
[13] Yusliza, M.-Y. et al. (2020). A structural model of the impact of green intellectual capital on sustainable performance. Journal of Cleaner Production, 249, 119334. http://dx.doi.org/10.1016/j.jclepro.2019.119334.

Ch. Rupa, Divya Midhun Chakkarvarthy

# Web-Based Knowledge Management Distributed Application for Medical Certificates Using Blockchain Technology

**Abstract:** Globally, disruptive technologies play a vital role in ordinary people's lives to technocrats, small-scale industry to multinational companies, and private sectors to government sectors. Providing security and privacy preservation of the sensitive data in various applications related to a bank, healthcare sectors, etc., are becoming significant issues with advanced technologies. Unauthorized persons and also attackers get benefits from the features. Permissionless and open access features will become advantages to the cyber attackers to crack the data. These issues will be resolved by adopting some knowledge management systems (KMSs) like blockchain technology, and Artificial Intelligence. Therefore, this chapter proposed a web-based knowledge management decentralized application using blockchain technology in healthcare sectors. Especially in the healthcare centers, people's sensitive personal data such as name, address, financial details, and cause maintain as records. Cyber attackers treat the medical records as treasure. Advanced featured technology, blockchain incorporation with web-based applications can reduce the attacks like ransomware, phishing, Brute force, steals credentials, and Man in the middle attack. The technology's main inherited features are distributed, immutable, transparency, security, and peer-to-peer network communication. This hybrid approach, a combination of Web 3.0 and blockchain technology, increases the privacy and preservation of official medical documents issued by healthcare centers. This chapter illustrated the process of a proposed decentralized application using Ethereum-based Blockchain with Metamask wallet. This work's main strength is the proposed application deployed results over an ethereum blockchain platform on test RPC, which are laps in the maximum related literature survey works currently available. Furthermore, presented comprehensively about the deployment cost of the application operations in terms of smart contracts and execution time for uploading credential requests.

**Keywords:** Web 3.0, Blockchain, Knowledge Management, Ethereum, Metamask, DApp, Official Health Documents

## 1 Introduction

The Internet has developed a lot over the last three decades. Web 3.0 is an advanced version of Internet services designed as a semantic web to create intelligent and open web applications connected [1]. It includes various technologies to create real-

https://doi.org/10.1515/9783110722789-009

time and user-friendly applications and make the Internet an intelligent information system [2]. In Web 3.0 technology, data is integrated in a decentralized way. Also, Web 3.0 applications use blockchain technology to make a secure database and verified communication. The advanced features of Artificial Intelligence, machine learning, data analytics, and blockchain technology such as intelligence, analytics, distributed, transparency, immutable, and security are accommodating to achieve the objectives of Web 3.0 [3].

According to WHO (World Health Organization), 95 percentage healthcare applications or centers were exploited by hackers or attackers in 2017. The primary threat to healthcare data security is ransomeware attacks [4]. Hackers tend to steal health records because they treat them as most valuable assets than physical currencies or assets. These medical records consist of a patient's sensitive information like a name, address, financial details, phone numbers, and social security members. Moreover, as per a report of Trustwave [5], medical records worth in the black market is very high. Each health data record is valued at up to $250. Key features of Web 3.0 and blockchain technology can reduce threats or attacks on medical records.

Blockchain technology and Web 3.0 are both known as distributed ledger technology (DLT). Blockchain technology-based browser enabled the features of Web 3.0, i.e., brave browser [6]. It fixed major undermining issues in earlier versions like Web 2.0 and Web 1.0 and resolved them by adopting blockchain technology features. Blockchain is peer-to-peer (P2P)-based interconnected blocks using crypt hash values. The significant concepts of blockchain technology are mining and consensus protocols.

According to National Information Center, India, Blockchain is a KMS [7]. Because it can create the transactions, store them in a block, and process the block over a blockchain. Committed transactions are immutable by connecting them with hash values of previous blocks. The processing of a blockchain depends on consensus protocols. These protocols are using to share, manage, and transfer the data among the nodes in a blockchain. Proof of Work (PoW), Proof of Stack (PoS), and Ripple, etc., are the most standard consensus protocols [8].

Different blockchains types are available to deploy the proposed applications. Those are public blockchain, private Blockchain, consortium blockchain, and hybrid blockchain. The common similarities among all the blockchains are that every blockchain has nodes operating on P2P networks [9]. Each node maintains an own ledger that is updated frequently. Furthermore, every node verifies, receives, and initiates the transactions and also creates the blocks. The comparison among different blockchain types has shown in Table 1.

This chapter consists of information about the proposed system, that is, a blockchain-based distributed application (DAPP), to securely create healthcare certificates and store them over an ethereum-based public blockchain [10]. To implement this application uses the etherum platform with Truffle suit and metamask wallet as mentioned in Table 1.

**Table 1:** Comparison of blockchain types.

| Property | Public | Private | Consortium/ federated | Hybrid |
|---|---|---|---|---|
| Access | Anyone | Single Authorized Organization | Multiple Organizations | Authoritative access allowed, but some elements are private |
| Consensus | Permissionless PoW, PoS | Byzantine Fault Tolerant (BFT), Permission | Permission | private and other public |
| Network | Decentralized | Partially decentralized | Decentralized | Partially decentralized |
| Platforms | Ethereum, NEO, Ripple, Bitcoin, Stellar, Litecoin | Hyper ledger, CORDA, Multichain, Quorum | Energy Web Foundation, R3 CORDA | Dragonchain |
| Smart contract (Languages) | Etherum: Solidity NEO: Java, Python, C#, VB.net | Hyper ledger: Go, Chaincode CORDA: Kotlin | Private | Public/ Private |
| Tools | Ethereum: Truffle, metamask, Ganache, Remix, Geth, Mist | Hyper ledger: Hyper ledger-compose, visual studio CORDA: Flow | Hyper Ledger | Public/ Private |
| Use cases | Any application | Financial, Bitcoin Supply Chain, Healthcare, Real Estate, Insurance, Govt. services. | Any application | Global Finance & Trade, Supply Chain, Banking, Enterprises. |

The remaining portion of the chapter formatted as follows. Section 2 has information about fundamental concepts of blockchain technology and Web 3.0. Related work on the proposed concept has been mentioned as a literature survey in Sections 2 and 4 has proposed system architecture with the step-by-step process of the proposed system smart contracts are as shown as algorithms. Evaluation results and analysis of the proposed system are described in Section 5. And also shows comparison of proposed work with the existing works by considering some features. Section 6 has the chapter conclusion and future scope.

# 2 Rationale

Availability of the information to the user level of expectations is becoming very difficult. Services are like privacy, connectivity, and accessibility of the data from everywhere is a complex task. But this can be resolved by using Web 3.0 and blockchain technology features. The main basic concepts and features of these technologies are as follows.

## 2.1 Web 3.0: preamble [11–14]

The main features of Web 3.0 are Knowledge management like Artificial Intelligence, Connectivity, Ubiquity, Semantic Web, and 3D Graphics. Web 3.0-based application content can access by the number of applications at any time from anyplace. These applications' scope is to search, generate, share, and analyze based on the meaning of the words instead of keywords or numbers. Web 3.0-based DAPPs as shown as follows.

**Applications:**

- **Brave:** It is a Web 3.0-based browser. Chrome, Internet Explorer, etc. are based on Web 2.0 Applications (Apps).
- **Stroj & IPFS:** Designed DAPPs using Web 3.0 features to store maximum amount of data. Like a Google drive application in Web 2.0.
- **Status:** This application is used to communicate with each other using messages. It was designed using Web 3.0. Like Whatsapp, Sharechat applications which are running on Web 2.0.
- **Experty:** This DAPP uses to make audio, and video call skype that was designed using Web 2.0.
- **Steemit & Aksasha:** These are Web 3.0-based social network applications like Facebook and Twitter those are applications of Web 2.0.
- **Ethlance:** It is a Web 3.0-based Remote Job Platform. This application has built totally on blockchain and using only cryptocurrencies for transactions.

## 2.2 Blockchain technology: preamble [15–20]

**Key Elements:** A blockchain is a group of different concepts and methods. Those are Cryptography, Hashing, P2P network, distributed ledger system, Proof of work, Merkle tree, blockchain network, block, etc. Blockchain technology's main features are transparency, distribution, immutable, security, P2P connectivity, and decentralized. It is complicated to hack where data was not maintained in a centralized server. The main concepts of blockchain technology are briefed as follows.

- **Cryptography:** Asymmetric key cryptography is used in the blockchain for providing security, authenticity, and integrity to the data and users.
- **Hashing:** It is a nonreversible process that is used for encrypting the data of a block.
- **Peer-to-peer network (P2P):** The nodes here are not connected to any central system. It is a secure decentralized and distributed network.
- **A distributed ledger system:** Each record information is available at each node of the P2P network rather than at a central authority node.
- **Proof of Work:** It is a process performed by the nodes in blockchain technology. Generally, nodes add a block into the blockchain permanently by solving a mathematical puzzle given a system.
- **Merkle Tree:** Merkle tree structure saves the block space by keeping a single hash value rather than taking all the transactions.
- **Blockchain Network:** It contains several connected nodes by following a P2P connection. Every node in the blockchain can do proof-of-work (PoW), verify the transactions, and process them.
- **A Block:** A block is like a ledger or collection of records.

**Components of a block:** A block in a blockchain has the following attributes: Block header, block size, transaction data, and transaction counter. Table 2 shows the details of the characteristics of the block [21].

**Table 2:** Fields of a block.

| Attributes/ fields | Block header (bytes) | Block size (bytes) | Transaction data (bytes) | Transaction counter (bytes) |
|---|---|---|---|---|
| **Filed Size** | 80 | 4 | 1–9 | Variable (Min 400) |
| **Depiction** | – Metadata<br>– It has a version, hash value, Earlier block hash value, timestamp, nonce, etc. | Size of a block | Data of all the Transactions | Total number of transactions in a specified block |

Table 3 gives detailed information about block header components such as block version, hash value, a hash value of the earlier block, Nonce, Timestamp, Merkle root, and difficulty. This information is known as metadata, giving a closer look at a block and the transactions in a blockchain [22].

**Table 3:** Detailed information of block fields.

| Attributes/ fields | Block version (bytes) | Earlier block hash (bytes) | Merkle root (bytes) | Timestamp (bytes) | Nonce | Difficulty |
|---|---|---|---|---|---|---|
| **Filed size** | 4 | 32 | 32 | 4 | 4 | 4 |
| **Depiction** | Version number uses to track software or protocol updates | A hash value of the earlier block | A hash value of the Merkle tree root | A time of block creation | It is a number that is used for the PoW process | It set for PoW algorithm |

# 3 Literature survey

Ryahei Banno et al. [23] proposed a SimBlock, which is a simulator based on blockchain. It impersonates the P2P network over a blockchain. The blockchain network is adaptable because of its architecture that has various nodes with different configurable parameters. There are no ways to validate or trace the behaviors of a blockchain. This technology adoption rate is increasing daily due to its advanced features like immutable, transparency, and maintains the data as distributed ledger manner.

Gaohe Li et al. [24] proposed a new method for verifying electronic invoices' authentication using blockchain technology. It solves the tax and financial departments' problems by analyzing the risks in a digital invoice system by processing computer images. Initially, the sum of image pixels and modulus values are taken from the image to verify whether the electronic invoice has been tampered with or not. The identification and tracking of invoices are made using OCR (Optical Character Recognition) technology to check whether it is authentic or fake. This system automatically identifies the invoice to avert one invoice from utilizing once more.

According to Junfeng Shi et al. [25], the most recent research is done on blockchain technology. Here proposed a life cycle model using blockchain technology for Internet-based financial products. It anticipates the products life cycle curve based on the power grids state. Here the life cycle curve is classified into three stages. It starts with the input period, then the maturity stage, and finally, the recession period. It has a high reference value which helps in developing power finance.

Daniel Tse et al. [26] proposed storing usernames and passwords in an encrypted format using blockchain technology. The users provide their details that have usernames and passwords to validate the information authentication system. Many information systems are available for accessing the data, hence unable to remember the users' credentials. Most people forget their usernames and passwords and also facing a struggle to retrieve them again. The proposed system is using blockchain technology for storing the user's credentials. Blockchain technology is immutable. The identity information stored would not be affected by any cyber attacks.

M. Fakhar Manesh et al. [27], provided an attempt to conduct a comprehensive, structured, and objective analysis of academic research in the Knowledge Management literature in the Industry 4.0 era. The study uses bibliometric evaluation to uncover the field's layout and determine the direction of future research. Later, a systematic literature review has performed about the cluster results. Literature review determines the outcome of each cluster's unique contribution. The results of this study show that quality assurance professionals need to contemplate, realize, and incorporate all aspects of "Industry 4.0" progress in their firm to reduce future adverse effects on the firm's business. The only drawback observed is the presence of a noticeable fragmentation of some keywords in some clusters.

Maurizio Atzori et al. [28] discussed the ways of Data Exploration provided by the emerging Web 3.0 Paradigm. The proposed work significantly focuses on how Web 3.0 is different from its predecessors by having Human-Computer Interaction, Semantics, Patterns, Data Exploration, and Recommender Systems. The proposed work promotes the dissipation of the best results in all the domains related to Data Exploration in the broadest perception.

N. Iqbal et al. [29] developed an approach RIVIMS using machine learning techniques and blockchain-based smart contracts. RIVIMS refers to a reliable, intelligent veterinary information management system. This method has two segments: a blockchain-based information management system and data analytics models. Hyperledger is used to develop the information management system in RIVIMS. Data analytics and predictive models are built on a permission blockchain framework. These models intend to research the patients' details to find out underlying styles and construct a sturdy prediction model based on the machine learning algorithms. According to the predicted results, veterinary clinics' management can improve its future business plans, such as future resources and staff needs. This system allows the clinic to retain its customers and gives an advantage over its competitors.

K. P. Yu et al. [30] proposed a blockchain-enhanced security access control scheme to address the semi-credibility nature of the IIoT (Industrial Internet of Things) cloud platforms, and blockchain implements authentication, public keys, features of the users, and revocation lists. Domain administrators are responsible for developing security and privacy policies for the domain and implementing encryption functionalities. Intermediate decryption parameters will be returned from servers only when the features are mapped by considering the user's ID not available in the revocation lists. This scheme helps us store securely, control, update, and delete information for business intelligence data to track and recover malicious users.

L. Da Xu et al. [31] proposed a work that inspects state-of-the-art IoT (Internet of Things) security using blockchain.This work focuses on addressing security functions, challenges, technologies, approaches, and related IoT scenarios. The proposed work utilizes blockchain keys like decentralization, consensus mechanism, data encryption, and smart contracts to build distributed IoT systems. Smart contracts are used for device registration, data storage, and service provisioning. By utilizing these features, it prevents potential attacks, as well as reduced transaction costs.

Shuchih E. Chang et al. [32] proposed an analytical review and systematic research on blockchain-based supply chains to facilitate the technology's research. This work presents the benefits and problems prevalent in the supply chain paradigm that uses blockchain. A total of 106 review articles were examined to conclude blockchain technology usage in supply-chains. Digitalization, transparency, collaboration, traceability, and stakeholder involvement are significant issues resolving blockchain-based platforms. The integration of blockchain technology into supply chains can provide better and efficient performance, increase security, ensure distribution of governance, and help process automation. This paper presents the summary of applications of blockchain. The social impacts of these technological applications and their technical adoption/diffusion are also highlighted in this paper.

Christos Karapapas et al. [33] presented how smart contracts and IPFS (Inter planetary file system) systems be used by criminals to launch ransomware attacks. The ransomware attacks have been a significant issue for Internet users in recent times. The proposed system consists of authors and affiliates, the authors create the ransomware and store them in the IPFS, and affiliates buy this ransomware through smart contracts based on Ethereum. Then the affiliates use this ransomware to infect victims. Solidity uses for writing smart contracts, and these contracts are composed of seven functions. The IPFS offers excellent flexibility and robustness, and with the advantage of the churn of IPFS, it was possible to offer secured services for affiliates. The proposed mechanism has enhanced security and is for fellows to register and get the ransomware. It is hard to get this mechanism offline, which is its greatest asset.

Ryuya Uda [34] proposed an efficient mechanism to prevent the falsification and fabrication of video recordings by surveillance cameras. Videos recorded are signed and chained on the blockchain in this mechanism, which provides data integrity. The author proposed creating trust chains based on some predefined conditions and distinguishing cameras. The signer of the trust chain got rewarded. This mechanism proves to be efficient and effective against the fabrication of video recordings and is also cost-efficient.

Le Jiang et al. [35] have proposed a blockchain technology-based framework using OSN (Open Science Network) and DOSN (Distributed Open Science Network). It combines both because online social networking sites are becoming more and more general in today's life. Still, people face privacy leakage problems because of the developers' approaches like centralized data management systems, and lack of security protocols. The proposed method resolves this issue. In this work, the authors used real-time data set to evaluate the proposed system performance. The system's advantage is to solve the users' privacy issue, and the drawback is that it is necessary to design complicated protocols to support routing carefully.

According to G. Zyskind et al. [36], there are raised surveillance incidents cases and exploitation of the security breaches. The attackers or unauthorized parties are misusing the collected data from the breaches. In this work, the authors described a mechanism to control user's data on their own using a decentralized approach. Here proposed a protocol to turn a blockchain hooked on a control manager with an additional feature such as automatic accessing. Here the transactions are not operated based on crypto balance. The advantage is that there is more security regarding personal data.

The main objective of this research proposed by Shu Yang et al. [37] was to improve traditional blockchain protocol's linear structure. The authors used DAG, that is, Directed Acyclic Graph to avoid limited throughput. Here the blocks have arranged based on the widths and levels that are generated by the CoDAG (Compact Directed Acyclic Graph) structure. CoDAG provides more security than traditional blockchain.

# 4 Blockchain and web-based use case: decentralized application for official medical documents

Generally, any KMSs are categorized into three types such as intelligent technique-based systems, knowledge work systems, and enterprise KMSs. A web-based KMS with the advanced technology blockchain improves the user experience to capture the details quickly. Moreover, this application or system can process, organize, and share the data with all the authorized nodes in the network efficiently by following some standard strategy.

## 4.1 Web-based knowledge management DAPP

Knowledge management is a process for making better data decisions, quality communications with the information, and improving data agility [38]. The main components of knowledge management are process, content, users, and strategy. Figure 1 shows knowledge management-based decentralized application architecture running using blockchain technology and Web 3.0.

In blockchain-based DAPP for official medical documents, users initially request a medical document through Web application to process the given details. Metamask, a cryptocurrency wallet, asks for confirmation to process the user's request over a blockchain. The process starts on a blockchain by following some stages whenever confirmation is received from the admin or physicians. Minors begin to mine the data by solving a mathematical puzzl e before making the data as a transaction in a block [39]. In the next stage, generates a Merkle root hash value to the data. The result stores as a transaction in a block created by a minor. Finally, the block is connected with the blockchain, as shown in Figure 1.

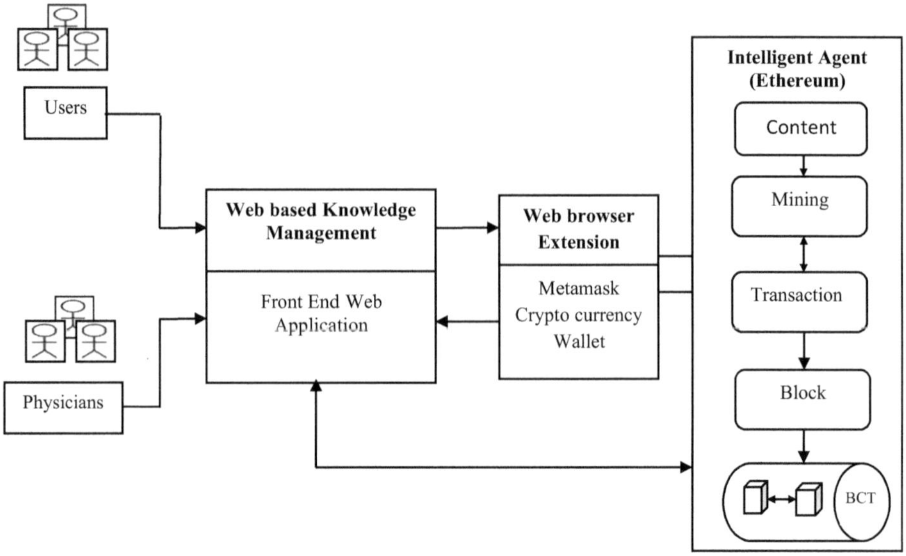


**Figure 1:** A KMS using Web and blockchain.

The step-by-step process of establishing a connection between a web-based application to a blockchain network to deploy a decentralized application is shown as Algorithm 1. As shown in Figure 1, initially, either user or admin from healthcare centers must enter their credentials for integrity checking to start the web application. Activate the Metamask cryptocurrency wallet that is a web browser extension, after completing the process of validating the operator's credentials.

Metamask acts as an interface between web applications to the blockchain network [40]. In the proposed system Ethereum-based public blockchain is using to compile and deploy the system operations. A Metamask wallet provides an Eth balance for performing the proposed system operations like creating, issuing, and verifying the medical documents on a blockchain network.

**Algorithm 1: Web-based DAPP**

**Input:** Credentials of the users or physicians or admins
**Output:** Connection establishment with a blockchain network

**Begin**
  **Repeat**
    **For** each user 'u'
      **If** ( Validate(u_ details) = = true)
        Activates web browser Extension Cryptocurrency Wallet.
        Invite for authentic confirmation.

```
                If confirmed
                        Connection established with Ethereum blockchain
                Else
                        Not connection with Blockchain network
                End If
        Else
                Not activated
        End if
    End for
  Until (activate all legitimate requests on BCT)
End
```

## 4.2 Smart contracts for official medical documents

In the blockchain environment, smart contracts are lines of code where solidity programming uses to write smart contracts for a specified system's required operations. The proposed method, maintenance of an official medical documents system, has two primary functions: issuing a user-desired medical document by authority and verifying the document by users, administrators, and physicians. Algorithm 2 shows the smart contract pseudocode of the operation to issue a medical certificate.

**Algorithm 2: Smart Contract to set the values for Issue_doc ( )**

**Input:** Required details i.e., cause, date, name, Health_center name, doc_name for a document
**Output:** issue an official document interms of an unique BCT-based ID.

```
pragma solidity >=0.4.22 <0.6.0; // It is compiler version
contract MedicalCert
Begin
    uint public Hospital_Reg_No;
  string public Hospital_Nmae; //bytes16
  string public Doctor_Name; //bytes16
  string public Certificate_Type;//bytes32
  string public Gender;
    string public cert_Holder_Name;
    string public date_of_issue;
    string public date_of_receipt;
  string public Receipent_Name;
    uint public UID_Reciepent;
    for all the users
```

```
        function set ( variables ) public
        Begin
            Set the values to all the variables.
            End
         Return (BCT_ID)
    End for
End Contract
```

As shown in a smart contract, first needs to mention the version of a compiler using the instruction “pregma solidity” where considered “≥ 0.4.22 < 0.6.0” is a version of the compiler to deploy the smart contract. Later needs to mention a name to the smart contract. Here we used “MedicalCert” as the name of the proposed smart contract. As shown in algorithm 2, the document required details collect from the user by declaring them in the solidity programming-based smart contract.

**Algorithm 3: Smart Contract to Verify_doc ( )**

**Input:** Document ID i.e BCT_ID, cert_Holder_Name,Certifictae_Type
**Output:** Existed or not

```
pragma solidity >=0.4.22 <0.6.0; // It is compiler version
contract MedicalCert
Begin
    Repeat
        for all the documents
            function verify ( variables )
            Begin
                If hash (Holder_Name (BCT_ID)) = = hash (Holder_Name)
                        && hash (certificate (BCT_ID) = = hash (certificate_Type)
                        Integrity check passed.
                        Else
                        Integrity check failed.
                End If
            End
        End for
    Until (validate all the documents)
End
```

Algorithm 3 shows smart contract pseudocode for the function integrity checking of the issued certificate or a medical document. In this process, verify the hash value of the received blockchain-based ID certificate with the existing documents’ hash values. Any hash value existed with the same document ID and with the same holder name considering it as a valid certificate. Here we used the Keccak-256 approach to evaluate the hash value of the required documents. Keccak 256 operates in the ethereum-based blockchain [41]. Keccak-256 in a consensus engine called Ethash. Keccak is a family of hash functions like SHA, which is used in the Bitcoin network.

# 5 Experimental evaluation and analysis

The proposed Web and blockchain-based DAPP for managing the official medical documents. That is implemented using Web3.js, Nodejs, Metamask, testrpc, and Ethereum-based blockchain on localhost 8545 test network. Figure 2 shows a web page for taking the user and admin credentials to operate on blockchain-based DAPP.

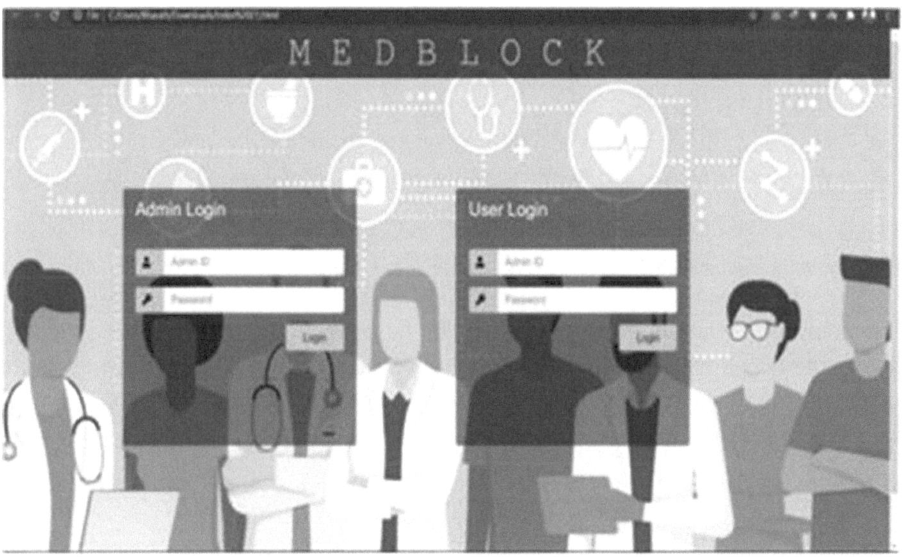


**Figure 2:** Web-based DAPP-related users credential screen.

Metamask Wallet asks confirmation to make integrity checking of the user's credentials. After it is confirmed, it operates by deducting some Gas (a cryptocurrency that measures in Eth) from the wallet. Figure 3 shows the fields of an official medical document the admin or physicians enter after confirming the users' details. This DAPP performs issuing a certificate and verifying the certificate operations on top of the credentials entered into the Web page, as shown in Figure 3.

Metamsk takes confirmation to perform any operation over a blockchain network. Figure 4 shows the process of the Metamask Wallet. Figure 4 (a) shows the initial amount in a Metamask wallet, i.e., 99.8313 ETH. Figure 4 (b) shows consumption cost, i.e., 0. 004774 ETH, and a confirmation request to perform "issue a certificate"

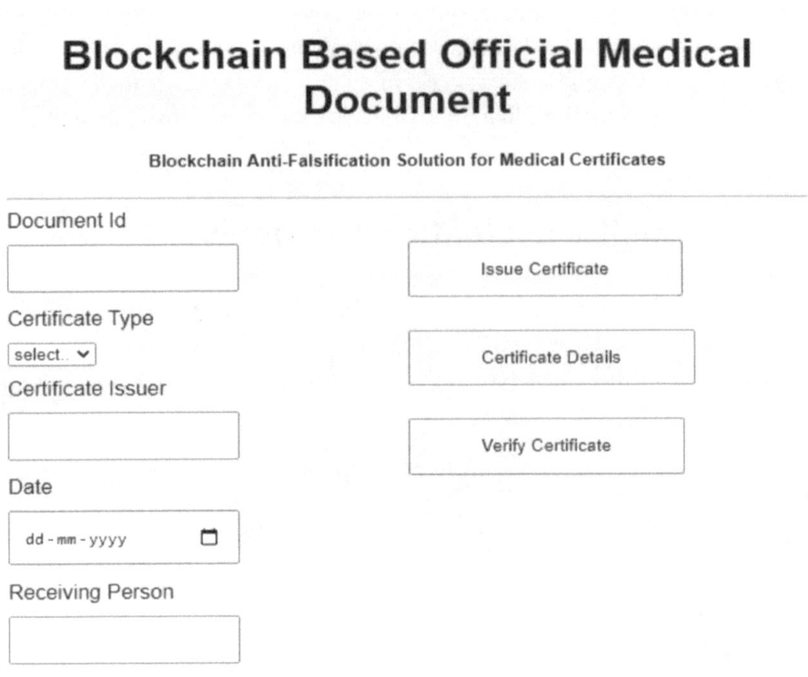


**Figure 3:** Web- and blockchain-based medical certificates creation.

operation for the admin or physician's credentials according to Figure 3. Figure 4 (c) shows the resultant Metamask wallet after confirmed the wallet request on the issue of a certificate. It shows the current available balance in the wallet after consumption for issue operation, i.e., 99.8281ETH.

## 5.1 Cost analysis

Each operation is executed on a public blockchain that is Ethereum based. Here the process refers to a transaction that takes some Gas in terms of cost to run over an Ethereum blockchain, where gas is a unit to operate. Ethereum gas paid here as Ether. Ether is known as crypto-fuel to deploy the applications over a blockchain network. A transaction cost includes the cost of execution and other expenses related to the deployment of a contract and data transfer into a blockchain. An execution cost includes smart contract internal storage and blockchain manipulation state.

Table 4 shows the costs of various operations over a blockchain network. Here, one Ether is $10^9$ gwei units. Hence consumption costs to perform the operations have shown using ETH and Gwei. The proposed DAPP smart contract has four functions:

**Figure 4:** Metamask Wallet shows an operation deployment cost.

enroll user or admin credentials, set the document credentials, issue an official document, and verify the document by all the authorized users.

**Table 4:** Smart contracts deployment costs.

| Operations | Cost (Eth) | Cost (Gwei) |
|---|---:|---:|
| User credentials | 157,412 | 0.000157412 |
| Set document credentials | 83,230 | 0.000083230 |
| Issue | 157,388 | 0.000157388 |
| Verify | 72,140 | 0.000072140 |

Table 5 shows the execution and transaction costs of the system operations. Generally, blockchain takes little time to read the data. Furthermore, transactions are validated by the blockchain nodes and make them as records into a block. Minors validated the block and appended it to the blockchain. The average time of the request and transaction processed over TestRpc-based localhost 8545 is not fixed. It varies based on the system configuration and size of the transactions.

**Table 5:** Execution and transaction cost analysis.

| Smart contract | Execution cost ( Gas) | Transaction cost (Gas) |
|---|---:|---:|
| User | 682,101 | 863,212 |
| Admin/authority | 1,587,652 | 2,165,431 |
| Request | 818,553 | 1,154,324 |
| Document | 832,233 | 975,438 |
| Share | 665,321 | 854,281 |

Figure 5 shows cost consumption by considering different "n" value: number of certificates maintenance over a blockchain. Here various cost factors represent the cost consumption for the credentials, such as Gas limit, Gas cost, Gas price in terms of CGWEI, and total Eth.

## 5.2 Security analysis

Analyzed the proposed Web and blockchain-based DAPP by considering two threat scenarios. Also compared with the existed systems by considering technical features and security services.

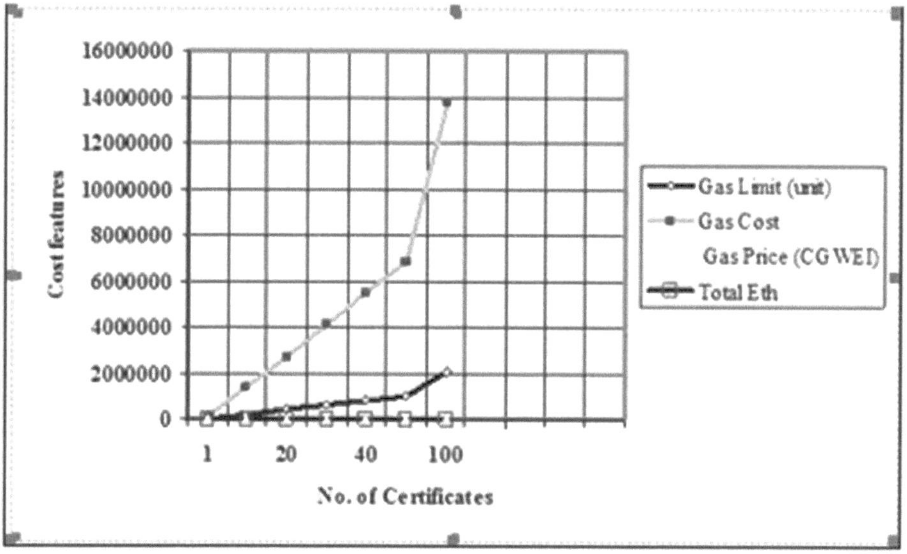


**Figure 5:** Gas consumption for different medical documents.

Theorem 1: Assume that any person wants to manipulate the data in the medical document illegally.

Proof: The proposed application uses blockchain technology to maintain and issue a unique ID for every medical document. The main feature of the blockchain is immutable. If the data entered into a blockchain, it could not trace or manipulate because of connecting the data with a nonreversible hashing method. Hence, it is very challenging for the attacker to control the medical document credentials.

Theorem 2: Assume that an adversary tried to send incorrect data into a block as a transaction for maintaining over a blockchain network.

Proof: To enroll the medical documents data over a blockchain network, uses intelligent web system in the proposed application. For that reason, validate the information by the authority before enrolling the details into a blockchain environment. There is no option to register the details by anyone other than an authorized authority. Therefore, incorrect information or fraud details enrollment can resist through the proposed supplication. Table 6 shows the comparison between proposed and existed systems.

**Table 6:** Comparison the proposed system performance with others.

| Features<br>Authors | Availability | Distributed | Admin | Authentication | Data privacy | Integrity |
|---|---|---|---|---|---|---|
| | No | No | No | Yes | Yes | Yes |
| | No | No | Yes | Yes | Yes | Yes |
| | No | Yes | No | Yes | Yes | Yes |
| | No | Yes | Yes | Yes | Yes | Yes |
| | Yes | Yes | No | No | Yes | Yes |
| Proposed system | Yes | Yes | Yes | Yes | Yes | Yes |

# 6 Conclusion and future scope

As per the fabric ventures report, Web 3.0 benefits more than 3 billion people every day with its objectives such as open, permissionless, distributed, and trustless. Web 3.0 includes the core layers of Web 2.0, such as Social, mobile, and a cloud, making it a disruptive technology. This work proposes a decentralized application using Web and blockchain technology to generate and preserve the official medical documents with security. Many critical challenges existed in the current system to issue and maintain medical records. Every medical certificate consists of the users' sensitive data that felt like a treasure to the attackers. This chapter mainly focused on trustworthy-based access controlling the data. Only admins and physicians can operate on the medical document rather than everyone. Here, the proposed system's performance is evaluated by considering the factors such as cost and security. This work can extend to using private blockchain in the future. And also, it needs to integrate the proposed application on real-time data sets.

# References

[1] Atzori, M. & Pes, B. (2019). "DEW 2019: Data Exploration in the Web 3.0 Age," 2019 IEEE 28th International Conference on Enabling Technologies: Infrastructure for Collaborative Enterprises (WETICE), Napoli, Italy, 138–139.

[2] Hatzivasilis, G. et al. (2018). "The Interoperability of Things: Interoperable solutions as an enabler for IoT and Web 3.0," 2018 IEEE 23rd International Workshop on Computer Aided Modeling and Design of Communication Links and Networks (CAMAD), Barcelona, Spain, 1–7.

[3] Ohei, K.N. & Brink, R. (2019). Web 3.0 and Web 2.0 technologies in higher educational institute: methodological concept towards a framework development for adoption. International Journal for Infonomics (IJI), 12(1), 1841–1853.

[4] AlSabeh, A., Safa, H., Bou-Harb, E. & Crichigno, J., (2020). "Exploiting ransomware paranoia for execution prevention," ICC 2020-2020 IEEE International Conference on Communications (ICC), Dublin, Ireland, 1–6.

[5] Alnabulsi, H., Islam, R. & Talukder, M. (2018). GMSA: Gathering multiple signatures approach to defend against code injection attacks. IEEE Access, 6, 77829–77840.

[6] Heitmann, N., Pirker, B., Park, S. & Chakraborty, S. (2020). "Towards building better mobile web browsers for ad blocking: the energy perspective (WiP Paper)" LCTES '20: The 21st ACM SIGPLAN/SIGBED Conference on Languages, Compilers, and Tools for Embedded Systems, 146–150

[7] Zhou, I., Makhdoom, I., Abolhasan, M., Lipman, J. & Shariati, N. (2019). "A Blockchain-based File-sharing System for Academic Paper Review," 2019 13th International Conference on Signal Processing and Communication Systems (ICSPCS), Gold Coast, QLD, Australia, 1–10.

[8] Zhang, S. & Lee, J.-H. (2020). Analysis of the main consensus protocols of blockchain. ICT Express, 6(2), 93–97.

[9] Shala, B., Trick, U., Lehmann, A., Ghita, B. & Shiaeles, S. (2020). Blockchain and trust for secure, end-user-based and decentralized iot service provision. IEEE Access, 8, 119961–119979.

[10] Priya, K.L.S. & Rupa, C. (2020). "Block chain technology based electoral franchise", IEEE International Conference on Innovative Mechanisms for Industry Applications (ICIMIA), Bangalore.

[11] Elberkawi, E.K., El-firjani, N.F.M., Maatuk, A.M. & Aljawarneh, S.A. (2016). "Usability evaluation of web-based systems: A new method and results", International conference on engineering & MIS (ICEMIS), Agadir, Morocco, 1–5.

[12] Elkobaisi, M.R., Maatuk, A.M. & Aljawarneh, S. (2015). A proposed method to recognize the research trends using web-based search engines. ACM International Conference Proceeding Series, 24, 192–203.

[13] Atzori, M. & Pes, B. (2018). "Summary report for the data exploration in the Web 3.0 Age (DEW) track", 27th IEEE International Conference on Enabling Technologies: Infrastructure for Collaborative Enterprises (WETICE), 139–141.

[14] Zareen, S.S., Guangmin, S., Qadri, S.U. & Qadir Khan, A. (2019). "Towards the development of framework for semantic web interface," 2019 IEEE 4th Advanced Information Technology, Electronic and Automation Control Conference (IAEAC), Chengdu, China, 2723–2727.

[15] Reimers, T., Leber, F. & Lechner, U. (2019). "Integration of blockchain and internet of things in a car supply chain," 2019 IEEE International Conference on Decentralized Applications and Infrastructures (DAPPCON), Newark, CA, USA, 146–151.

[16] Gao, J. et al. (2019). "Towards automated testing of blockchain-based decentralized applications," 2019 IEEE/ACM 27th International Conference on Program Comprehension (ICPC), Montreal, QC, Canada.

[17] Dedeoglu, V., Jurdak, R., Putra, G.D., Dorri, A. & Kanhere, S.S. (2019). "A trust architecture for blockchain in IoT", Proc. ACM Int. Conf. Mobile Ubiquitous Syst. Comput. Netw. Services (MobiQuitous), 190–199.

[18] Al-Jaroodi, J. & Mohamed, N. (2019). Blockchain in industries: a survey. IEEE Access, 7.

[19] Tanwar, S., Bhatia, Q., Patel, P., Kumari, A., Singh, P.K. & Hong, W.-C. (2020). Machine learning adoption in blockchain-based smart applications: the challenges, and a way forward. IEEE Access, 8, 474–488.

[20] Rupa, C. (2020). Security and privacy of UAV data using blockchain technology. Journal of Information Security and Applications, Elsevier.

[21] Halpin, H. & Piekarska, M. (2017). "Introduction to security and privacy on the blockchain", IEEE European Symposium on Security and Privacy Workshops (EuroS&PW).
[22] Mishra, R.A., Kalla, A., Singh, N.A. & Liyanage, M. (2020). "Implementation and analysis of blockchain based dapp for secure sharing of students' credentials," 2020 IEEE 17th Annual Consumer Communications & Networking Conference (CCNC), Las Vegas, NV, USA, 1–2.
[23] RyoheiBanno, K. (2019). "Simulating a blockchain network with simblock" 2019 IEEE International Conference on Blockchain and Cryptocurrency (ICBC).
[24] Li, G., Sun, S. & Li, X. (2020). "Block chain digital invoice system risk assessment and system development research", 2020 International Conference on Intelligent Transportation, Big Data & Smart City (ICITBS), 724–727.
[25] Shi, J., Zhu, Q. & Jing, N. (2017). "Research on life cycle of power financial products based on block chain factor", 2017 4th International Conference on Industrial Economics System and Industrial Security Engineering (IEIS), 1–5.
[26] Tse, D., Huang, K., Cai, B. & Liang, K. (2018). "Robust Password-keeping System Using Block-chain Technology," IEEE International Conference on Industrial Engineering and Engineering Management (IEEM), Bangkok, Thailand, 1221–1225.
[27] Fakhar Manesh, M., Pellegrini, M.M., Marzi, G. & Dabic, M. (2021). Knowledge management in the fourth industrial revolution: mapping the literature and scoping future avenues. IEEE Transactions on Engineering Management.
[28] Atzori, M., Koutrika, G., Pes, B. & Tanca, L. (2020). Data Exploration in the Web 3.0 Age, Future Generation Computer Systems.
[29] Iqbal, N., Jamil, F., Ahmad, S. & Kim, D. (2021). A novel blockchain-based integrity and reliable veterinary clinic information management system using predictive analytics for provisioning of quality health services. IEEE Access.
[30] Yu, K.P., Tan, L., Aloqaily, M., Yang, H. & Jararweh, Y. (2021). Blockchain-enhanced data sharing with traceable and direct revocation in IoT. IEEE Transactions on Industrial Informatics.
[31] Da Xu, L., Lu, Y. & Li, L. (2021). Embedding blockchain technology into IoT for security: a survey. IEEE Internet of Things Journal.
[32] Chang, S.E. & Chen, Y. (2020). When blockchain meets supply chain: a systematic literature review on current development and potential applications. IEEE Access.
[33] Karapapas, C., Pittaras, I., Fotiou, N. & Polyzos, G.C. (2020). "Ransomware as a service using smart contracts and IPFS", IEEE International Conference on Blockchain andCryptocurrency.
[34] Uda, R. "Data protection method with blockchain against fabrication of video by surveillance cameras", ICBCT'20: Proceedings of the 2020 The 2nd International Conference on Blockchain Technology, ACM, March 2020.
[35] Jiang, L. & Zhang, X. (2019). BCOSN: a blockchain-based decentralized online social network. IEEE Transactions on Computational Social Systems, 6(6).
[36] Zyskind, G. & Nathan, O. et al., "Decentralizing privacy: Using blockchain to protect personal data", Security and Privacy Workshops (SPW) IEEE, pp. 180–184, 2019.
[37] Yang, S., Chen, Z., Cui, L., Xu, M., Ming, Z. & Xu, K., "CoDAG: an efficient and compacted dag-based blockchain protocol", 2019 IEEE International Conference on Blockchain (Blockchain), 2019, pp. 314–318.
[38] Fonou-Dombeu, J.V., "Ranking semantic web ontologies with ELECTRE," 2019 International Conference on Advances in Big Data, Computing and Data Communication Systems (icABCD), Winterton, South Africa, 2019, pp. 1–6.
[39] Faour, N., "Transparent E-Voting DApp Based on waves blockchain and RIDE language," 2019 XVI International Symposium "Problems of Redundancy in Information and Control Systems" (REDUNDANCY), Moscow, Russia, 2019, pp. 219–223.

[40] Taş, R. & Tanrıöver, Ö.Ö., "Building a decentralized application on the ethereum blockchain," 2019 3rd International Symposium on Multidisciplinary Studies and Innovative Technologies (ISMSIT), Ankara, Turkey, 2019, pp. 1–4.
[41] Khatoon, A. (2020). A blockchain-based smart contract system for healthcare management. Advances in Blockchain and Distributed Ledger Technology (DLT) for Industry 4.0 Technologies, 9(1).

Deepa Sahai Garg

# Incorporation of Indian Companies Using Integrated Web Forms as Ease of Doing Business Initiative: An Overview

**Abstract:** India is a leading performer in the international business setup, and its economic growth and development has made it a destination for investment for many global investors. With simplification and rationalization of regulatory framework as well as usage of information technology, it is becoming one of the fastest-growing technologically progressive knowledge economies in the global emerging markets. The knowledge-based economy has become a vital part of production, savings, investment, and consumption function which is based upon important resources like innovative ideas and intellectual capital [15]. Knowledge and knowledge management are becoming an important economic and organizational resource. The businesses have been facilitated and have grown over the last decade using Information and Communication Technologies (ICT) and various knowledge products to enable the borderless online transactions and have better global presence [23]. Corporate, legal, business and technological reforms initiated by the government have a positive impact on easing the business regulations from registration up till winding up of companies and thereby boosting investment inflows within the country.

With the intelligent web entering its third generation and offering features of semantic web, enhanced web browsing, improved information linking, artificial intelligence, machine learning, better web technologies, and resourceful searching from worldwide databases, there is a paradigm change in the world of internet and the way businesses are run, marketed, and set up are being redefined. The evolution made from Web 1.0 (read-only websites) to Web 2.0 (read-write websites) and now Web 3.0, also called the web of everything, allows user interaction on the web pages with increased connectivity in between the data. Web 3.0 makes the process of online applications more simpler and user friendly [38]. Taking advantage of this technological revolution, several business reforms have been initiated to ease the procedures relating the starting of business.

One of the important corporate reforms which has been undertaken is web-based Simplified Proforma for Incorporating Company Electronically (SPICe) Plus form as a part of Ease of Doing Business initiatives on improving the business climate for the registration of companies. The incorporation of the company form of organization by means of the advanced features of artificial intelligence using web-based filing of the application and the electronic documents and various regulatory filings have become totally paperless and are completely online on the Ministry of Corporate Affairs (MCA) web portal. Starting a business has been made lot easier,

https://doi.org/10.1515/9783110722789-010

seamless through the online incorporation procedure using web-based form, that is, SPICe+ along with electronic documents like e-memorandum of association and e-articles of association [30].

This chapter elaborates and discusses the improvements brought about by the information technology revolution and its significance for the business growth and entrepreneurial development. Various initiatives undertaken by government in implementing single e-compliance window mechanism, the greater ease of starting business, major ease of doing business factors relating to the information technology solutions, important corporate reforms, the impact of newer web technologies in terms of opportunities, and challenges that such technologies will present to the corporate world are also presented in the chapter.

**Keywords:** Web 3.0, Knowledge Management, Ease of Doing Business, Regulation, Technology, SPICe Plus form

# 1 Introduction

Any economy which is advanced and modern necessarily needs to be knowledge-driven that ensures balanced achievement of socioeconomic goals and investment growth. Indian businesses operate under globally competitive environment toward sustaining their business growth and as an attractive investment economy. The entrepreneurs in India have always struggled to start business due to burdensome policies, procedures, and regulations for the businesses. After the 1991 reforms, various changes brought regulatory changes to reduce difficulties the businesses face and providing the conducive environment. The regulatory reforms in the form of entry liberalization and privatization have had a significant impact lowering entry barriers and spurred investment in the long run, which is an important engine of economy's growth [7]. Governments in various economies have stringent regulations that burden entrepreneurs. Such regulations not only limit entrepreneurs' ability to operate a private business with less barriers but the entrepreneurs also start doing informal activity to avoid regulations and manipulating the laws, taking employment in other countries. Also, the investors usually do not find such economies that use excessive regulations, a destination for the investment [16]. Establishing or incorporating a manufacturing or service enterprise in India requires numerous procedural compliances, regulations, and permissions. India has approximately 12 lakh active companies. Various entry restrictions and barriers like difficulties in the start-up procedures and higher cost of doing business affect economic growth and thereby the growth of new businesses is hindered [23]. India's regulatory procedural mechanisms are being reformed to be given new and easy forms primarily to enable the businesses to adapt to and to integrate with the international global standards and thereby attract foreign investments and encourage entrepreneurship in nation building. Efficient administration and a simple regulatory environment are highly valued by foreign investors while investing in the economies. An enabling regulatory environment provides easy entry and exit of enterprises.

The government's Ease of Doing Business (EODB) initiative has been supportive with a noteworthy role, and the government has given foremost importance to it. Indian state governments and Union Territories have also launched several business reforms under the Business Reform Action Plan (BRAP) to provide the businesses with enabling an enterprise-friendly regulatory environment and provide the real-time ranking depending upon the reforms undertaken [10]. Ease of doing business is crucial for the entrepreneurs while setting up new businesses, large/medium/small in terms of regulatory procedures particularly relating to purchase of land, process involved in getting the property registered, and single window mechanism for obtaining the clearances along with a single online application for obtaining clearances under different acts.

The time factor, procedural formalities, regulatory hurdles, documentation requirements, and the broader restrictions for the manufacturing and service sector enterprises in the incorporation process have been simplified and made easier. Many regulatory changes have been introduced over the past years like the Limited Liability Partnership Act 2008, the Companies Act 2013, the Insolvency and Bankruptcy Code 2016, and the Goods and services Tax Act 2017, along with various procedural changes and reforms These initiatives have simplified the existing regulations along with the introduction of Information and Communication Technology (ICT) to make governance and their application more well organized and effective. At the same time improved digitization of businesses along with the integration of various approvals obtained from different agencies has further reduced the number of steps of incorporating enterprise. Various business reforms undertaken have started reflecting the results. Electronic governance and technological advancements are playing a crucial role in persuading a culture of better corporate compliances and thus achieving higher and easier levels of ease of doing business. The increased usage of information technology tools assists in making the process of approval fast, effective cost wise, and completed in a time-bound manner and hence improving and facilitating the business.

# 2 Methodology

This chapter is of review nature exploring the research area and is based on secondary data sources. The data about the indicators for the ease of doing business, current scenario, issues, and problems have been taken from publications from Ministry of corporate affairs, PHD Chamber of Commerce and Industry of India, Niti Aayog, DIPP (Department of Industry Policy and Promotion), Department of Promotion of Industry and Internal Trade (DPIIT), Doing Business Report of the World Bank, different national and international publications, newspaper articles collected from various websites, and e-resources relating to research study.

## 3 Literature review

The review of the recent conceptual literature suggests that there has been implementation of the latest ICT-based applications in the corporate and business world to streamline the administrative and regulatory procedures related to businesses in India showing positive results.

Technological advancements in the regulatory environment and improvements in local business conditions are happening at a fast pace in many emerging markets [5]. Singh and Jaiswal [1] in their paper suggested that implementing and employing e-governance and technology will lead to filing the procedures electronically in online mode and thereby no visits to the government offices shall be required for obtaining clearances and thus reduction in the cases of bribery and corruption. The Indian government has persistently applied latest technology and web tools to streamline processes to remove the corruption practices by the business community in doing business [36]. Advanced technology should be used for ease of doing business for the enterprises to grow with minimum human interface especially in the government services and using AI should be the focus for government interface [4].

R. K. Yadav et al. [27] in their paper studied the impact of e-governance and Internet readiness on the ease of doing business. A supportive business environment promotes competition, encourages innovation, and also promotes entrepreneurship in the economy, thereby leading to economic growth [21]. Improved digital entrepreneurship is of high relevance as technological developments create many new opportunities for the emerging entrepreneurs [33]. The government that cares about its economy's growth, well-being, and development pays more attention to friendlier laws, administrative procedures, and regulations for local small- and medium-size enterprises. Djankov et al. [32] in their paper had differentiated countries on the basis of how the countries regulate the entry of new businesses and emphasized upon the relationship between excessive regulation and its impact as the hindrance to entrepreneurship.

The governments can only function with the support of administrative procedures for collecting information like income tax, and for market activities being regulated to implement public policies. The administrative procedures worldwide seem to be increasing as governments require more information for their activities [26].

L. Alfaro and A. Chari [20] examined the effects the license raj reforms had in India on the size of the firm and the allocation of resources. The authors noted that with the easier start up rules, the number of small firms increases with increase in their performance and productivity. A reduction in resource allocation distortions over the same period and deregulation will lead to more firms and less incumbent power [3].

Simple business regulations provide the small firms the opportunity to grow, innovate, and thereby come into formal sector from the informal sector of any economy [9]. The ease by which entrepreneurs can start a business could be a strong determinant whether investors will start a business in the first instance [35]. J. Estevao et al. [19] in their paper concluded that assessing credit alone is not the major factor for knowing the suitability of any business, as there are many difficulties which are significant while dealing with the regulatory authorities. Better and minimum regulatory measures are needed, and their efficacy keeps varying over time. A supportive business environment enables the businesses to grow more and also contribute more toward the growth of the economy. The more successful the private sector is, more businesses are created, more shall be the opportunities for employment generation [35]. T. Canare [35] in their paper suggested that overall ease of doing business has a positive impact on business creation which is most strongly driven by the starting a business. In addition, the effect of the starting a business component is driven by the financial cost rather than the time and administrative cost. Finally, results change when the analysis is applied to non-high-income countries only. As per the Ease of Doing Business, 2020 report there is a significant difference between low-income and high-income economies for the ease of starting a business [7].

Goel et al. [24] in their study concluded that the technology and regulations play an important role for the country in inviting more capital flows and thereby enable more growth and development in the financial sector. Doing Business Report, an annually published study, contains the names of 10 countries that improved the most on the ease of doing business after a detailed analysis of costs and implementation of regulatory reforms and procedures in developing economies every year from 2003 onward in 190 countries, and ranks them. All the economies which are scoring high in the Doing Business Report has shown results as better entrepreneurship giving rise to better employment opportunities, improved incomes, and thereby increased taxation revenues. Also, the incidence of corruption and bribery has been seen to be reduced with the adoption of electronic means of compliance for the regulatory approvals and requirements [9]. The EODB index measured through the Doing Business Report has been recognized as an important measure among countries and as one of the important knowledge products helping in building the structure of varied regulatory reforms in several developing countries.

## 4 Ease of Doing Business (EODB): an introduction

Since the year 2004, the World Bank in its Doing Business Report ranks 190 countries for the ease in doing business based on several criteria called EODB ranking. The report consists of 294 business regulatory reforms spread across the 10 areas measured by doing business [28]. These reforms addressed various segments that are: starting or incorporating business, obtaining construction, or building permits, electricity connection, registering and transferring property, access to credit, protecting the interest of minority investors, payment of taxes, international trading across the borders, enforcing contracts, and resolving insolvency proceedings. The economies are scored based on their performance in 10 EODB segments (12 areas from the year 2020) included in the ease of doing business ranking to find the relative ease in their business environment [7, 9]. The EODB index is a ranking system established by the World Bank group and the ranking is based upon the simple and streamlined procedural regulations in an economy. The lower the value of EODB index, the higher is the rank for the country, indicating simpler and conducive regulatory environment for setting up businesses in that country [8]. Doing business ranking is considered an important index which is used by global investors to evaluate the investment proposals in various countries. One of the crucial Ease of Doing Business indicators is ease of starting a business which is an important indicator while assessing the Ease of Doing Business index. This indicator ranks based on all the procedures required to start a business and operate the business and the time and cost involved to complete the compliance procedures. With respect to starting a business, Doing Business Report defines various stages for the smooth establishment and conduct of any business. These stages start with formalities related to incorporating a business and employing workers, to getting the location for the business and then obtaining electricity connection, registering property, and dealing with the construction permissions [6]. Accessing finance, easy dealing with daily operations, and conveniently operating in a secure business environment are some of the crucial factors looked at while evaluating any economy.

## 5 Ease of starting business in India after technological and corporate reforms for online incorporation of companies and limited liability partnerships (LLPs)

Various economies which score high on the EODB rankings have commonalities like the usage of electronic systems for online procedures to make online incorporation a seamless process through a single window or one agency rather than multiple agencies. The top 20 ranking economies, as per the Doing Business Report, all have the facility of online incorporation, electronic portal for tax filing platforms, and online procedures for property transfers. They also have sound and transparent business regulations. Moreover, 11 economies have online procedures for permissions related to construction [7]. India has undertaken various digital efforts toward making ease of starting business seamless and to simplify the process of starting a new business. As per Doing Business Report 2020, India is among the 10 top performers

**Table 1:** Ease of Doing Business indicators.

| Indicator set | What the indicator measures |
|---|---|
| 1. Starting a business | The time, cost, and procedural steps along with paid-in minimum capital requirements to start a limited liability company, simplified preregistration and registration formalities, simplified post registration procedures, improved online procedures. |
| 2. Dealing with construction permits | The time, cost, and streamlined procedural steps involved to fulfil the formalities regarding warehouse, time taken for processing the applications, effective quality control systems and safety nets in the system relating to construction permitting, streamlined new building regulations, improved electronic platforms or online services. |
| 3. Getting electricity | The time, cost, and procedural steps to obtain electrical connection for a new warehouse, the consistency of the electricity supply, and the transparency related to tariffs, time in obtaining a new connection, online software to process applications for a connection, streamlined process of approvals through digitization processes, cost of new connections. |
| 4. Registering property | Streamlining property registration process, the time, cost, and procedural steps related to registering, transparency of information and transferring a property and the quality of the land administration system taxes or fees and administrative efficiency. |
| 5. Getting credit | Strengthening the rights of borrowers and lenders through collateral laws, various credit information systems, and their depth. Improving the sharing of credit information. |
| 6. Protecting minority investors | Strengthening minority investor's protection through ease of shareholder suits, expanded shareholders' role in company management, minority shareholders' rights and increased disclosure requirements for the related-party transactions, extent of director liability. |

**Table 1** (continued)

| Indicator set | What the indicator measures |
|---|---|
| 7. Paying taxes | Making it easier to pay taxes through enhanced online and electronic system for filing and payment of taxes, total number of taxes paid, payments, time involved, and total tax and post filing processes, improved tax audit processes. |
| 8. Trading across borders | Making it simple to trade across the borders through improved electronic submission and processing of documents for import and exports, time and cost required to export the product and to import, number of documents necessary to export and import, strengthening border infrastructure for imports and exports. |
| 9. Enforcing contracts | Making it simplified to enforce contracts, introducing changes to the civil rules, electronic case management systems, electronic filings, cost and time involved in resolving a dispute and enforcing a contract, specialized commercial courts, alternative dispute resolution framework introduced. |
| 10. Resolving insolvency | Making it simple to resolve insolvency, cost, time, and recovery rate for insolvency proceedings, new restructuring procedures, strengthening creditor's rights. |
| 11. Employing workers | Flexibility in employment regulation. |
| 12. Contracting with the government | Procedures and time to participate in and win a works contract through public procurement and the public procurement regulatory framework. |

and is moving closer to the international best practices on most of the indicators and implemented a one fifth of all regulatory reforms in the year 2018–19 showing impressive progression in its ranking. Over the past many decades, various initiations which had been on back seat have taken huge leaps with the technological revolution. These advancements in technologies have more effect on the ease of doing business than the number of reforms undertaken over the years in terms of liberalizing the economy.

India's scores have improved in seven out of ten indicators and are moving closer to the international practices. India's ranking has gone up from the 142nd position in 2014 to the 63rd in 2019 in Ease of Doing Business rankings out of 190 countries as per the World Bank's Doing Business Report [11]. India has shown an improvement in the indicators – starting a business, dealing with construction permits, trading across borders, with the user-friendly electronic business atmosphere and the decline in the timeline [12]. Also, the wide gap between regulations and its actual implementation has reduced. To achieve this, the Ministry of Corporate Affairs has taken great initiatives for providing "Ease of starting Business" over the last few years which include recently introduced web form SPICe+ and various efforts to reduce the number of steps required for establishing business. Out of the indicators in the EODB report, there are three indicators like starting business, protecting the interest of minority, and resolving insolvency of the companies, which come in the broader regulation of Ministry of Corporate Affairs [22]. The Ministry of Corporate Affairs has been significantly contributing toward the technological improvements especially with respect to starting a business and reducing the number of procedures thereby providing enabling business environment [18].

On an average, across the countries that are part of the ranking, it takes seven procedures with twenty-five days for starting a business. When we see in New Zealand, it takes only one procedure, half of a day and almost zero fees in New Zealand for starting a business, whereas an entrepreneur must wait for 208 days in Suriname and over 144 days in Republican Bolivariana de Venezuela. As per the Doing Business Report 2020, on an average 18 days are now needed to set up any business in India and reducing it further to five days is an important agenda of government of India [7].

India started by establishing an online Value Added Tax (VAT) registration and later replaced it with the faster GST (Goods and Services Tax) registration, e-stamping, reduction in the registration fees, elimination of the minimum capital requirement, merging the applications for the Permanent Account Number (PAN), and the Tax Account Number (TAN), and enhanced the online application system features, fully integrating multiple application forms into a single incorporation form. The various other measures introduced in this respect are the removal of the minimum paid-up capital requirements, reduction in the number of procedures/steps involved, and cost for a small- to medium-sized limited liability company to start up. Various parameters like starting a new business, obtaining construction permits, and resolution of the insolvent businesses have shown noteworthy development leading to considerable improvement in the business environment in the country leading the way for the country offering global manufacturing hub to attract domestic and foreign investments. The Ministry of Corporate Affairs has contributed significantly toward the improvement of ranking in ease of starting a business like reducing the number of procedures, days, cost, and no minimum paid-up capital requirement for a company, a single form for multiple registrations, and so on. Following are some of the steps taken by the MCA as a part of the central government's ease of doing business initiatives:

- The minimum paid-up capital requirement for the private companies has been removed under the Companies (Amendment) Act 2015 by the Ministry of Corporate Affairs as a step toward increasing ease of setting up companies [22].
- Under the Government Process Re-engineering (GPR) initiatives, whereby for making the incorporation process easy, smooth, seamless, and reducing the number of

procedures required for starting a business, the Ministry of Corporate Affairs (MCA) had introduced "RUN – Reserve Unique Name" web service for name reservation in the year 2018. With the launch of SPICE+ and AGILE PRO forms, the RUN web service is available only for changing the existing company's name from February 15, 2020 onward [2]. Name Availability Rules have been simplified by MCA avoiding any ambiguity in the name reservation. The rate of name rejection has dropped, and the time taken for obtaining the approval has been reduced resulting in speed, transparency, and uniformity [22].

– Formation of Central Registration Centre (CRC) done under section 396 of the Companies Act, 2013, as an initiative under the GPR for the prompt incorporation-related services through a single online process. The establishment of CRC has been to bring the incorporation procedures parallel with the international best practices [18]. This was achieved by the launch of the SPICe and now SPICe+ form through a single integrated web form filed on MCA21 Portal. This web form provides a single window for ten different registrations or services by three central government ministries: Ministry of Corporate Affairs, Ministry of Labour, and Ministry of Finance and the State Government of Maharashtra, thereby relieving the companies from obtaining certifications under different ministries. SPICe+ tries to achieve onscreen filing and real time data validation. SPICe+ form has introduced a new user-friendly dashboard for accessing all the applications for SPICe+ and other linked forms and accessing already saved information related to SPICe+. It has two parts: Part A and Part B. Part A consists of name reservation only for new companies [2],[22].

**Table 2:** Part B of SPICe+ offers nine services under a single window [29,34].

| S. no. | Services covered |
|---|---|
| 1 | Incorporation of company |
| 2 | Allotment of Director Identification Number |
| 3 | Issue of PAN |
| 4 | Issue of TAN (Tax Deduction and Collection Account Number) |
| 5 | EPFO (Employees' Provident Fund Organization) |
| 6 | ESIC (Employees' State Insurance Corporation) registration |
| 7 | Profession tax registration (Maharashtra) w.e.f. February 23, 2020 |
| 8 | Open bank account |
| 9 | GSTIN (Goods and Services Taxpayer Identification Number) |

The company incorporated through SPICe+ form has to submit an application for a bank account by using a linked form: AGILE-PRO. The link to AGILE-PRO is enabled after part A SPICe+ is filled. AGILE-PRO web form has been made mandatory in all the cases. SPICe+ application cannot be submitted without this form. Many banks like Punjab National Bank, State Bank of India, Bank of Baroda, ICICI Bank, Kotak Mahindra Bank, and HDFC Bank have integrated with SPICe+. AGILE-PRO (replacing the old AGILE) is a part of SPICe +, under which application is made for Goods and Services Tax Identification Number (GSTIN), Employees' State Insurance Corporation (ESIC) registration, Employees' Provident Fund (EPF) registration, Profession Tax Registration (Maharashtra) and opening account in bank [14]. With the SPICe+ incorporation form, the number of procedures has been further reduced and will surely help improve the EODB ranking of India in future.

- The requirement of a making separate application for the allotment of Director Identification Number (DIN) for first-time directors of companies has been dispensed. The bank number is provided immediately now after incorporation, resulting in saving of days to start business. Integration of the registration for GST, EPFO, ESIC, and profession tax (for Maharashtra) along with company incorporation form is another milestone achieved. Before the introduction of SPICe+ form, application for GST registration, Employees State Insurance (ESI), and EPF registration were optional along with the company incorporation in SPICe form. But now all these registrations are mandatory for a company [22,29]. Shram Suvidha portal provides for the online single window registration and licensing for various stakeholders under the Ministry of Labour and Employment. Registration of companies under the EPF and ESIC are done on the MCA's portal through Spice+ and AGILE-PRO e-forms only at the incorporation time and not on Shram Suvidha Portal without any physical documents and physical visit. With this development the incorporation procedure has become faster. Though the introduction of this new step would require amendments under the respective EPF Act and the ESIC Act.
- The high regulatory, very high registration fees, and start-up costs borne by the private sector make it difficult to start business and young entrepreneurs who are bootstrapped do not venture and fear incorporating new companies. Starting a business in India has become less costly due to filing or registration fees getting eliminated owing to SPICe+ company incorporation form, electronic memorandum, and articles of association. Only stamp duty charges are to be paid for the incorporation by companies and the registration fees to be charged by MCA for all incorporations with authorized capital up to INR 15,00,000 have been eliminated. As per Doing Business Report 2020, India has made starting a business enterprise less costly by eliminating filing fees for the SPICe Plus company incorporation form and for electronic memorandum of association and articles of association [18].

- The Insolvency and Bankruptcy Code 2016, the first corporate solvency legislation, also introduced the Corporate Insolvency Resolution Process (CIRP) to provide for a time-bound insolvency resolution enabling ease of resolving insolvency [18]. IBC or the modern insolvency regime has provided a mean to form comprehensive strategy for the companies with effective tool to be able to reestablish financial capability and the creditors having better tools to negotiate and realize their money. The overall recovery rate from the creditors has increased, making India the best performer in South Asia based upon this indicator in the EODB index.
- Under the Companies Act 2013, making a company seal or the company's rubber stamp is no longer a legal requirement [22].

# 6 Ease of online incorporation of limited liability partnerships

Limited liability partnerships (LLPs) have become an important corporate entity for starting business in India, enabling and combining professional expertise along with entrepreneurial initiative. LLP is chosen due to the simple formation process and less legal formalities. MCA has introduced another GPR by introducing web service RUN-LLP (Reserve Unique Name – Limited Liability Partnership) (replacing Form 1 for reservation/change of name) for the reservation of the name and an integrated form FiLLiP (Form for incorporation of Limited Liability Partnership) (replacing Form 2), and the process has been centralized making incorporation of LLP a complete online process. FiLLiP is the single integrated application for name reservation, incorporation, and/or application for DIN/DPIN (Designated Partners' Identification Number) [22].

With these reforms and technological advancements being introduced, entrepreneurs will find ease in starting businesses in India as a part of ease of doing business.

# 7 India's commitment to business and technological reforms at the state level through Business Reform Action Plan

To bring competitiveness, efficiency along with transparency in the regulatory system, over the last many years, the Indian government has been taking actions to improve upon the regulatory climate for business operations. To provide right momentum and impetus toward the implementation of business reforms across various states and union territories, steps toward the comprehensive efforts and partnership between the central and state governments have been initiated which have facilitated toward bringing conducive and encouraging business regulatory environment for the businesses for promoting industrial growth and thereby creating better entrepreneurship opportunities.

State governments' role in the countrywide business reform exercise boosts the investors' confidence. On the similar lines as the World Bank's Doing Business EODB indicator's score and assessment, DIPP's initiative is to spread reforms beyond Delhi and Mumbai (the cities considered for doing business assessment). There was a need to broaden the base from these two cities as scored by the World Bank. The Department for Promotion of Industry and Internal Trade (DPIIT), the Ministry of Commerce and Industry, Government of India, introduced the BRAP. Ranking of states based on the implementation of BRAP started in the year 2014. Under BRAP, all the states and union territories of India are ranked pan India dependent on the business reforms undertaken on the basis of a set of chosen parameters on the similar lines of the Doing Business Report. India is among few nations having state-specific rankings, thereby helping the nation to improve its ranking. This ranking is done with the purpose of creating enabling and conducive regulatory business environment by reducing the regulations and thereby grading the states based on various measures of EODB. The DPIIT is responsible for the formulation and application of various measures relating to the development and growth of the industrial sector within the broader economic objectives. The various measures undertaken desires simplifying and digitizing the regulatory procedures/approvals, reducing the physical interaction as much as possible with the government officials, eliminating the discretionary processes, and expediting time-bound processes and other procedures as mentioned in BRAP recommendations. The Ease of Doing Business state rankings reflect the state's efforts to encourage competition and cut down on transaction and compliance costs. The BRAP has helped the states to enhance speed along with more transparency in the administrative processes and rationalizes the regulatory framework [17].

The DPIIT shares State Reforms Action Plan with States/Union territories to improve EODB. The Action Plans are spread across various reform areas which follow sector-specific approach to ensure coverage of all sectors so as to create enabling business environment across various sectors for setting up and for operating business in the state. Broadly the sector-specific reforms relate to making available information through Standard Operating Procedures to public, bringing policy and procedural modifications relating to amendments to Act and rules through various government orders, circulars, and so on by different departments and online system offering services by different departments with minimal physical interface and integration of services through single window portal. These rankings represent the ease of doing business in the states ranging from showing better transparency, increased efficiency, effective regulatory functions for the business setups focusing on low-cost and quick single desk approvals, the development of online systems, and regulating inspections covering all the aspects of the business [31]. According to the Department of Industrial Policy and Promotion, Government of India, under the State Reform Action Plan, 2019, Andhra Pradesh, Uttar Pradesh, Telangana, Madhya Pradesh, and Jharkhand are the top five states where business can be started and carried on smoothly. While in other states like Arunachal Pradesh, Meghalaya, Sikkim, Tripura, Nagaland, Odisha, and Manipur, entrepreneurs and businesses find difficulty in starting and operating business. As of March 2021, 20 states in India have successfully undertaken all the major business reforms for the ease of doing business, making them qualified for additional borrowing of 0.25 percent of the gross state domestic product. These 20 states have become eligible to raise Rs. 39,521 crores from the open market borrowings, as permitted by the Department of Expenditure. The developments due to the EODB reforms by the states will lead to faster growth of the states and the economy [39].

# 8 Single e-compliance window

To bring down the duplication of data filing by the businesses, the Ministry of Corporate Affairs (MCA), DPIIT, Securities and Exchange Board of India (SEBI), and the Reserve Bank of India (RBI) together are considering the proposal to collaborate for developing a single e-compliance window as the online platform without the need for the physical visit to several departments, with respect to application forms which share common data resources. The major goal of the proposed move is integrating the databases of all the ministries with which businesses have to register and reducing the compliance burden. With this platform, all the companies can upload the applications only at one portal and all the regulators shall access the data from that single repository. Single online compliance window system is a key reform point in the state action plan [25]. The features which are envisaged in this system are online submission, payment of application fee, following up of application status, downloading the certificate, and third-party verification. Such single repository can fulfil 80–90% requirements and any other additional requirement can be accessed through additional information requested separately. Registration and renewal of the licenses, approvals of permissions under various acts like the Factories Act 1948, the Boilers Act 1923, the Contracts Labour (Regulation and Abolition) Act 1970, the Shop and Establishment Act, the Building and Other Construction Workers (Regulation of Employment and Conditions of Service) Act 1996, the Inter State Migrant Workmen (RE&CS) Act 1979, Consent under the Air (Prevention and Control of Pollution) Act, 1981) &Water (Prevention and Control of Pollution) Act 1974, Authorization under the Hazardous and Other Wastes (Management and Transboundary Movement) Rules 2016, Registration/Renewal under Plastic Waste (Management and Handling) Rules 2011, the E-waste (Management and Handling)

Rules 2011, and so on. The dashboard regularly updates highlighting the number of registrations and renewals and the average time taken in obtaining the necessary approvals.

Various other reform areas or the investment enablers as mentioned in the 301 reforms of BRAP 2020 are as follows:

- Designing and implementing an online wizard/system where a user can enter in details like industry type, number of employees, location, and so on and can access the information about the approvals applicable for pre-establishment and pre-starting operations for the industrial unit. The online wizard shall also have access to online application forms link for various licenses/NOCs and the notifications. It shall provide the details regarding the fees, time taken, documents required, departments involved for the required NOCs or registrations, and so on, leading to accessibility to information in transparent manner.
- Legislative enactments ensuring the time bound delivery to the industries within the state, failing which, the violators shall face punitive action.
- State investment facilitation center or the promotion agency for investment promotion, underlining working procedures, service timelines, publishing online dashboard for grievance handling and updating regularly for the resolutions [31].

# 9 Conclusion and the challenges ahead

This chapter outlined the change brought about by the information technology revolution and its significance for the business growth and entrepreneurial development through the government encouragement and adaptation in the form of introducing technological-backed reforms. These included amendments and legislative reforms, tax reforms, banking reforms, and relaxed economic and trade policies which help the economy to grow. India is undergoing a period of unparalleled technological and economic advancement. Indian businesses have taken the advantage of the knowledge resources to upgrade their growth chart owing to the technological revolution using human capital facilitating the innovation and entrepreneurship and moving more closer to the international practices.

India is also moving swiftly toward digitization, and more and more government departments are deploying electronic platforms for public–government interactions, filing, and so on. For India, progress on the ease of doing business is a vital achievement as we seek greater global investments for lifting hundreds of millions to improved standards of living. As is observed in the World Bank report, economies that have adopted electronic means of compliance with regulatory requirements – such as obtaining licenses and paying taxes – experience a lower incidence of bribery, regulatory inefficiency, and red tape.

The regulatory landscape regime is continuously evolving. Various initiatives have been undertaken toward "Ease of Starting Business" by the Ministry of Corporate Affairs over the years for benefiting a large number of stakeholders. Latest web technologies have provided a variety of applications and digital tools on web for the business organizations, one of which is the SPICe+ web form for ease of incorporating companies. The creation of this form for the company registration through the agile and robust web platform of information will lead to smarter user experiences saving on time and efforts. Various emerging technologies like artificial intelligence, internet of things, machine learning, and 5G will further enhance the ease of doing business.

Harmonization of regulations with the global benchmarks results in reduction of regulatory procedures and thereby reduction in the compliance requirements. Rules and regulatory processes and their enforcement should be simple, prompt, effective, real, friendly, effective as well as emphatic to make India competitive in the global regulatory setting. The support from technological advancements will increase the swiftness of ease of incorporation and will further persuade the law makers and law enforcement agencies to bring changes and at the same time address to the challenges thrown by the adoption of technology and artificial intelligence at the workplace. For greater ease of doing business, consideration also has to be given for the impact of advancing technologies as well as the challenges and options that such technologies will present to the corporate, business, and legal ecosystem [25].

Various procedures that used to take many days can now be completed easily with the help of IT and technology solutions. With the Internet of Things (IoT), Web 3.0, artificial intelligence, machine-to-machine learning, and 5G several procedures will get easily connected and completed over the Internet; enabling and letting millions of applications to be handled and processed with the help of devices that can configure themselves, gather and process data with the autonomous devices, monitor usage, self-diagnose, and self-rectify faults, among other things. The Industrial IoT has opened various new possibilities for communications based upon intelligent and self-directed devices and bringing in industrial revolution entirely based on data [13].

The growth and development boosters in Indian economy like beneficial demographic conditions, talented human capital, strong start-up culture, increasing productivity, supportive government policy initiatives and a preferred investment destination for the foreign investors has helped to build a positive growth momentum. The policy initiatives are providing motivation and impetus to enable a friendly business environment and can make a visible difference in many critical areas too. The government has remained focused on enhancing the situation of the manufacturing sector in the country through its 'Make in India' project launched in the year 2014 as a program aiming to transform India into a global center for manufacturing and becoming important part of the global supply chain. All these efforts will make doing business simpler for start-ups in India and will establish an ecosystem to boost growth of start-ups and also create an enabling framework for receiving foreign investment, simplifying and cutting out the tedious paperwork/documentation and reporting procedures making India more investor friendly.

Information Technology and ease of doing business leads to transparent business processes and increased business growth. But adoption of technology may raise various issues regarding real-time performance, information privacy, security challenges that arise due to sensitive and confidential information, internal security, health hazards, environmental sustainability, which are also to be considered. Also, there are still many areas where there is immense scope for further improvement. Indeed, great progress has been achieved and but still much more has to be done. Despite the government's push, it is the receptiveness and efficiency of the administrative system that shall hold the key to the success. Technology can lead to growth and improvement in investment atmosphere only when the benefits accruing from technology are properly integrated into the conventional system of effective economic governance with the help of knowledge products and resources.

## References

[1] Singh, A. & Jaiswal, K.K. (2018). Ease of doing business in India: a vision of make in India. Economic Affairs. 63(1), 129–135. doi: 10.30954/0424-2513.2018.00150.16.

[2] "All About New MCA Filing Form SPICe+ (SPICe Plus) for Company Registration." https://blog.saginfotech.com/spice-plus-form-mca (accessed Nov. 11, 2020).

[3] Alesina, A., Nicoletti, G. & Schiantarelli, F. (2005). Regulation and investment. Journal of the European Economic Association. 3(4), 791–825. doi: 10.1162/1542476054430834.

[4] "Digital India: Ease of doing business, AI should be focus areas for Digital India: Nandan Nilekani – The Economic Times." https://economictimes.indiatimes.com/tech/internet/ease-of-doing-business-ai-should-be-focus-areas-for-digital-india-nandan-nilekani/articleshow/76730193.cms

[5] "Doing Business 2011: Making a Difference for Entrepreneurs." https://www.worldbank.org/en/news/press-release/2010/11/03/doing-business-2011-making-difference-entrepreneurs (accessed Dec. 23, 2020).

[6] "Doing Business 2020." https://www.doingbusiness.org/en/reports/global-reports/doing-business-2020 (accessed Dec. 19, 2020).

[7] "Doing Business 2020: Comparing Business Regulation in 190 Economies. Washington, DC: World Bank," pp 1–149, 2020. [Online]. Available: http://documents1.worldbank.org/curated/en/688761571934946384/pdf/Doing-Business-2020-Comparing-Business-Regulation-in-190-Economies.pdf.

[8] "Doing Business 2020: Reforms Boost India's Business Climate Rankings; Among Top Ten Improvers for Third Straight Year." https://www.worldbank.org/en/news/press-release/2019/10/24/doing-business-india-top-10-improver-business-climate-ranking (Accessed Nov. 11, 2020).

[9] "Doing Business. (2019): Training for Reform, Comparing Business Regulation for Domestic Firms In 190 Economies," pp 1–311, Accessed: Dec. 11, 2020. [Online]. Available: https://www.doingbusiness.org/content/dam/doingBusiness/media/Annual-Reports/English/DB2019-report_web-version.pdf.

[10] "Ease of Doing Business for MSMEs." [Online]. Available: www.commercedu.com. (accessed Dec. 18, 2020)

[11] "Ease of doing business in India: Myths and realities | Deccan Herald." https://www.deccanherald.com/opinion/panorama/ease-of-doing-business-in-india-myths-and-realities-793919.html (accessed Dec. 18, 2020).

[12] "Economic Survey, (2020–21)." https://www.indiabudget.gov.in/economicsurvey/doc/echapter_vol2.pdf.

[13] Sisinni, E., Saifullah, A., Han, S., Jennehag, U. & Gidlund, M. (2018). "Industrial Internet of Things: Challenges, Opportunities, and Directions; Industrial Internet of Things: Challenges, Opportunities, and Directions," doi: 10.1109/TII.2018.2852491.

[14] "Form SPICe+ (SPICe Plus) for Company Incorporation." https://taxguru.in/company-law/form-spice-plus-company-incorporation.html (accessed Nov. 9, 2020).

[15] Kefela, G.T. (2010). Knowledge-based economy and society has become a vital commodity to countries. International NGO Journal. 5(7), 160–166. Accessed: Nov. 11, 2020. [Online]. Available: http://www.academicjournals.org/INGOJ.

[16] Assaad, H., Celaya, H., Cruikshank, E. & Foran, J. (2011). "Investing for The Future Emerging Markets Private Equity Opportunity," http://avanzcapital.com/resources/Whitepapers/em_pe_aug_2011.pdf (accessed Dec. 26, 2020).

[17] "Improving Ease of Doing Business for MSMEs: A Review of Union and State Initiatives in India." Accessed: Dec. 26, 2020. [Online]. Available: http://ficci.in/spdocument/20765/Knowledge-Paper_Improving-EODB.pdf.

[18] "Improving the Ease of Doing Business in India: India's Recent Initiatives." https://www.researchgate.net/publication/338736858_Improving_the_Ease_of_Doing_Business_in_India_India%27s_Recent_Initiatives.

[19] Estevão, J., Lopes, J.D., Penela, D. & Soares, J.M. The doing business ranking and the GDP. A qualitative study. Journal of Business Research. 115, 435–442. Jul. 2020, doi: 10.1016/j.jbusres.2019.11.067.

[20] Alfaro, L. & Chari, A. (2014). Deregulation, misallocation, and size: Evidence from India. Journal of Law and Economics. 57(4), 897–936. doi: 10.1086/680930.

[21] Balcerowicz, L. & Rzońca, A. "Puzzles of Economic Growth Public Sector Governance." Accessed: Dec. 12, 2020. [Online], doi:10.1596/978-1-4648-0325-3.

[22] "Ministry of Corporate Affairs – Major Reforms & Initiatives." https://www.mca.gov.in/MinistryV2/reforms.html (accessed Dec. 22, 2020).

[23] "Moving Toward Knowledge-Based Economies: Asian Experiences." Available: https://www.adb.org/sites/default/files/publication/29699/knowledge-based-economies.pdf. Accessed: Dec. 14, 2020. [Online].

[24] Goel, N. & Singh, G. (2018, Jul.). The role of technology and regulations in capital flow to India. Australasian Accounting, Business and Finance Journal. 12(2), 87–103. doi: 10.14453/aabfj.v12i2.6.

[25] NITI Aayog, "Ease of Doing Business; an Enterprise Survey of Indian States," 2017. Accessed: Dec. 15, 2020. [Online]. Available: https://www.idfcinstitute.org/site/assets/files/12781/easeofdoingbusiness_report.pdf.

[26] "Reducing Administrative Barriers to Investment: Lessons Learned," 2006. http://regulatoryreform.com/report/reducing-administrative-barriers-investment-lessons-learned-2006/ (accessed Dec. 11, 2020).

[27] Yadav, R.K., Bagga, T., Sinha, A. & Kaul, S. A study on awareness of E-governance and its impact on ease of doing business. International Journal of Advanced Science and Technology. 29(9), 4306–4315. May 2020, Accessed: Dec. 21, 2020. [Online]. Available: http://sersc.org/journals/index.php/IJAST/article/view/16808.

[28] Mendoza, R., Canare, T. & Ang, A. "Doing business: A review of literature and its role in APEC 2015," Ateneo School of Government Faculty Publications, Jul. 2015, Accessed: Dec. 16, 2020. [Online]. Available: https://archium.ateneo.edu/asog-pubs/31.

[29] "SPICe is now SPICe+ with 10 Services Combined | LegalWiz.in." https://www.legalwiz.in/blog/spice-form-company-registration-with-10-varied-services (Accessed: Nov. 21, 2020). [Online].

[30] "SPICe+ Web Form filing process." https://ecpl.live/icai/26022020/SPICe_plus_Key_features.pdf (accessed Dec. 9, 2020).

[31] "State Reforms Action Plan 2020," pp 1–41. https://eodb.dipp.gov.in/PublicDoc/Download/30336 (accessed Feb. 12, 2021).

[32] Djankov, S., La Porta, R., Lopez-De-Silanes, F. & Shleifer, A. "Quarterly Journal Of Economics The Regulation Of Entry," vol. 1, pp 1–37, 2002. Accessed: Dec. 11, 2020. [Online]. Available: https://scholar.harvard.edu/files/shleifer/files/reg_entry.pdf.

[33] Kraus, S., Palmer, C., Kailer, N., Kallinger, F.L. & Spitzer, J. Digital entrepreneurship: A research agenda on new business models for the twenty-first century. International Journal of Entrepreneurial Behaviour and Research. 25(2), Emerald Group Publishing Ltd., 353–375, Feb. 21, 2019, doi: 10.1108/IJEBR-06-2018-0425.

[34] Maheshwari, S. "Incorporation of Company -Form SPICE+." Accessed: Nov. 9, 2020. [Online]. Available: https://taxguru.in/company-law/incorporation-company-form-spice-plus.html.

[35] Canare, T. "The Effect of Ease of Doing Business on Firm Creation," pp 555–584, 2018. Accessed: Dec. 5, 2020. [Online]. Available: https://ideas.repec.org/a/cuf/journl/y2018v19i2canare.html.

[36] "The Journal for Governance Professionals Ease of Doing Business-India's Commitment to Reforms," pp 1–176, 2020. [Online]. Available: http://support.icsi.edu.

[37] "The road to 50 expediting business regulatory reforms in India." https://www.phdcci.in/wp-content/uploads/2018/10/EoDB-Report-by-PHDCCI-and-KPMG.pdf (accessed Nov. 29, 2020).

[38] "What is Web 3.0? The Evolution of the Internet – Blockgeeks." https://blockgeeks.com/guides/web-3-0/ (accessed Dec. 14, 2020).

[39] "20 states complete ease of doing business reforms, eligible for additional borrowing – The Economic Times." https://economictimes.indiatimes.com/news/economy/policy/20-states-complete-ease-of-doing-business-reforms-eligible-for-additional-borrowing/articleshow/81604739.cms.

Tarakeshwar Gupta, Anuradha R. Tiwari

# A Study on Currently Practiced Effective Training Methods from a Knowledge Worker's Perspective in Outsourcing Services Sector

**Abstract:** This book chapter delves into the various training methods used in practice to address the various training objectives or training needs identified in the outsourcing services sector comprising BPO and KPO (business process outsourcing and knowledge process outsourcing). The employees who conduct or are associated to conducting these training programs are knowledge workers (KWs) such as trainers, learning, and development professionals, and managers. These KWs play a crucial role in their respective organizations in the outsourcing services sector and often help, train, equip, mentor, employees who lack skills required to execute a certain task. KWs in outsourcing services sector play the most crucial roles as they often have to implement policies and act as a critical communication point between the senior and junior managers. However, this group is often neglected when it comes to their own development. Middle managers often lack planning/project/financial management skills which are very critical to implementation of policy. Many a times, they have the knowhow of the job (the technical aspects of it) but are woefully short of people management skills, problem solving skills, interpersonal skills, and so on, which eventually are a large part of their KRAs (key responsibility area) now. Hence, it becomes imperative to equip them with pertinent skills. Middle management or managers are a part of KWs and hence training these knowledge workers with required skill becomes very important. Training methodology includes methods, materials, techniques, and resources used to implement the training or workshop and transfer new knowledge, skills, and attitudes to participants. The primary objective of this book chapter is to explore the latest trends in training methods in the outsourcing services sector from a KW's perspective in India. The chapter will also explore the effectiveness of these training methods in terms of achieving their training goals/ training objectives with respect to training needs. These training needs are identified through a questionnaire survey from a KW's perspective. Some of the training objectives identified in outsourcing services sector are acquiring new knowledge or skill, change in attitude, problem-solving skills, enhancement of interpersonal skills, and retention of knowledge and to address these training needs the training methods are lecture method, programmed instruction method, case study method, sensitivity training method, role-playing method, business games method, movie films method, conference (discussion) method, and television lecture method. For this exploratory study, data was collected in the form of questionnaire from knowledge workers who are responsible for capability building or boosting intellectual capital in the outsourcing services sector. Questionnaires emailed to KWs and were 76; filled were 60, which is a

https://doi.org/10.1515/9783110722789-011

return rate of 78.94%. Snowball technique has been used to collect samples. The findings illustrate extensively used effective training methods such as role-playing training method followed by programmed instruction method, and case study training method. Since Web 3.0 has come up as a technically advanced medium that allows users to read, write, and execute and furthermore it enables machines to perform which earlier was expected from humans only; hence, it has the potential to play a huge role in the future to be used a technology in training of knowledge workers. The chapter also focuses on usage of Web 3.0 in training of knowledge workers in the outsourcing services sector which can further enhance capability building or boosting intellectual capital.

**Keywords:** Training Methods, Effective Web 3.0, ITeS, Outsourcing, Knowledge Worker (KWs)

# 1 Introduction

Knowledge Workers (KWs) in the outsourcing services sector such as trainers, learning and development professionals, and managers play crucial roles. These KWs also help in upskilling employees who lack skills needed to perform a certain task. The KWs also have to frequently enforce policies and serve as the key point of contact between senior and junior managers. However, this group is often overlooked when it comes to their own development. Middle managers also lack skills in planning/project/financial management skills that are very important to policy implementation. They often have the knowledge of the job (the technical aspects of it) but are utterly devoid skills pertaining to managing people, problem-solving abilities, and interpersonal skills, and so on, which are now a large part of their KRAs (key responsibility areas). Therefore, equipping them appropriate skills is crucial. Middle management or managers are essentially a part of KWs.

Training is an important element in creating a high-work performance culture [1]. Training is used to enhance knowledge, skill, and attitude deficiencies. According to Mandakini [2], training methodology includes methods, materials, techniques, and resources used to implement the training or workshop and transfer new knowledge, skills, and attitudes to participants. Before selecting the training methodology, training objectives or training need identification (TNI) needs to be done to address the knowledge, skill, and attitude deficiencies.

## 2 Research objectives

The primary objective of this research-based book chapter is to explore the latest trends in training methods in the outsourcing services sector from a KWs perspective in India. The book chapter will also explore the effectiveness of these training methods in terms of achieving their training goals/training objectives with respect to training needs. These training needs are identified through a questionnaire survey from a KW's perspective. Also, it will explore the usage of Web 3.0 in training of knowledge workers.

## 3 Literature review

Most companies decide to outsource their operational activities such as customer services, HR, payment services, insurance claims, administration, and so on to another organization who in turn have developed core competency in these areas; such organizations are outsourcing services companies. Individuals working in such organizations are KWs as they exhibit a certain amount of knowledge to execute their tasks.

The literature review summary is shown in Table 1.

**Table 1:** Literature review summary table.

| Terms | Sources |
|---|---|
| Acquiring new knowledge or skill | Skinner [12–14] |
| Adult learning | Maria & Elena [5] |
| Business games method | Garris et al. [67]; Faria et al. [66]; Lopes et al. [64]; Faria [65]; Faria et al. [66] |
| Change in attitude | Janis & King [22]; King & Janis [23]; Culbertson [21]; Festinger & Carlsmith [19]; Harvey & Beverly [20]; Janis & Mann [24]; Petty, R. E., & Brinol, P. [18] |
| Conference (discussion) method | Johnson & Smith [83]; Costin [81] |
| Enhancement of interpersonal skills/ people management skills | Maier [33]; Van Schacck [32]; Maier & Maier [36]; Bolda & Lawshe [31]; Campbell & Dunnette [30]; Maier & Hoffman [34, 35] |
| Knowledge worker | Drucker [3]; Alvesson [4] |
| Learning | Logan et al. [7] |
| Lecture method | Carroll et al. [10]; Reith [45]; Mazur [42] |

**Table 1** (continued)

| Terms | Sources |
|---|---|
| Movie films method | Paivio et al. [78]; Richardson & Kile [73]; LeFevre, [70]; Seago [74]; Sherin [75]; Wang & Hartley [76]; Brophy [69]; Moreno & Valdez [71]; Borko & Pittman [68]; Pryor & Bitter [72] |
| Problem solving/troubleshooting skills | Parnes & Meadow [27]; Cohen et al. [28]; Fox [26] |
| Programmed instruction method | Mahoney et al. [49]; Weiss & Green [47]; Carroll et al. [10]; Skinner [11, 14–16]; Ayutthaya & Koomsap [46] |
| Retention of knowledge | VanderMeer [41]; Dietrick [38]; Sodnovitch & Pophorn [40]; Verner & Dickinson [39]; Bennett & Rebello [37] |
| Role-playing method | Buchan [55]; Harris [56]; Hershey [57]; Ments [60]; Hughes et al. [61]; Johnson & Johnson [54]; Sogunro [53]; Cranton [58, 59] |
| Sensitivity training method | Seashore [52]; Hinrichs [51] |
| Spaced learning | Sutton et al. [9]; Kolb & Kolb [8] |
| Television lecture method | Sherry [85]; McIsaac & Guanwardena [86]; Moore & Lockee [87] |
| Training | Goguelin [17]; Heathfield [1] |
| Training methodology | Mandakini [2] |
| Transfer of hard skills and soft skills | Arthur [6] |

Knowledge workers will be specialized. Applied knowledge is effective when it is specialized. This demands an ability to communicate ideas {3}. Formal education, analytical and theoretical knowledge, and a habit of lifelong learning are prerequisites [3]. Earlier, the industrial worker needed the capitalist but now an organization needs the knowledge workers [3]. This study is focuses on outsourcing services sector. Work done by KWs is perceived as very qualified, demanding, and intellectually stimulating [4].

Training needs to be devised in a manner that learning occurs. Learning will occur if one of the following takes place: (a) Attitude should change. (b) Knowledge should increase. (c) New skills acquired. KWs working in Outsourcing services are working adults. Adults as learners have specific characteristics according to Maria and Elena (2008) [5], such as (1) Adults participate in the learning process with specific intents, goals, and expectations. (2) Adults already have certain knowledge and experience as well as fixed perspective. (3) Adults have already developed personal styles of learning. (4) Adults are bound to self-directed activities throughout their lives. (5) Adults have to deal with certain obstacles in their learning process. (6) Spaced learning method is an effective adult learning method.

The training method adopted plays a significant impact on the transfer of both hard skills and soft skills [6]. Various studies have discovered that spaced learning over time helps people learn more quickly and remember better, and it has been found to be very effective in various areas, from sales training to language learning to medicines [7]. Training or a learning program with spaced learning in mind will educate the learners with a concept or learning objectives and then allow a period of time (days, weeks, or months) for the trainees to experiment what they have learnt [8]. "Spaced learning" is more superior that "massed learning" [9].

# 4 Conceptual framework

Different training methodologies are used to impart learning of KWs in outsourcing services sector. These training methods are specific to training objectives or the training needs identified. A training method is one of the factors that generally determine how effective a training program is. Some of the training methods used to enhance learning of KWs in outsourcing organizations are: demonstration, discussion, debates, lectures, presentation, role-play, game activities, case studies, electronic media usage (TV, videos, and films), and so on.

# 5 Training objective

The primary objective of training is expected measurable result that trainees or learners would reach after they have completed their training. Some of the training objectives identified in our purview of our study as per literature review are shown in Table 2.

**Table 2:** Training objectives identified as per literature review.

| Sl. no. | Training objectives |
|---|---|
| 1 | Acquiring new knowledge or skill |
| 2 | Change in attitude |
| 3 | Enhancement of interpersonal skills/ people management skills |
| 4 | Problem solving skills/troubleshooting skills |
| 5 | Knowledge retention |

These training objectives are similar to skills used in outsourcing organizations which is also endorsed by the KWs as per the data collated through questionnaire.

## 5.1 Acquiring new knowledge or skill

Acquiring new knowledge or skill is the process of processing and storing new information in memory, whose effectiveness is often measured by how well the information can be extracted from memory at a later stage. Acquiring new knowledge or skill is one of the training objectives predominantly seen in outsourcing organizations. The most effective training method for knowledge gain is programmed instruction method, and the least effective is lecture method [10]. In the programmed instruction method, the trainer controls the interaction of trainee's behavior with specific features of the instructional setting. It was through B. F. Skinner's [12–14] work that the programmed instruction method. Skinner's main focus was to state that self-education should be in the framework of reinforcement theory through programmed means. Training is a method of creating an urge to put into application what a trainee has learned, inspiring, motivating, and ensuring at the same time both the means and the appetite for using the knowledge that has been acquired. While teaching is restricted to giving knowledge based on reason about a given number of explicit ideas [17]. Some prominent researchers, both old and modern, who formulated a number of theories pertaining to training are Old Researchers: William James, J. F. Herbart, Wilhelm Wundt. Modern Researchers: E. L. Thorndike, E. R. Guthrie, C. L. Hull, B. F. Skinner, W. Kohler, and E. C. Tollermann.

## 5.2 Change in attitude

The word "attitude" is used to denote to a person's overall assessment of individuals, objects, and issues. Therefore, a person's attitude refers to individuals, objects, and issues positively or negatively or favorably or unfavorably in general [18]. Training programs pertaining to attitude is conducted across various sectors, and is very much relevant to outsourcing industry. Sensitivity training method is the most effective training method for change in attitude; however, role-playing, conference method, and case study are also effective in changing attitudes. Numerous studies conducted among students depict that the role-playing method is effective in addressing change in attitudes [19, 20]. Role-playing is effective if the participants in the role-playing exercise are asked to take a viewpoint opposite to their own and to verbalize this opposite viewpoint to others [21–24].

## 5.3 Problem-solving/troubleshooting skills

Problem-solving skill is a combination of problem recognition, problem analysis, strategy creation, information management, resource allocation, tracking, and problem assessment. One can also go back and forth through the steps, change the order when required, or even skip steps when it appears appropriate [25]. The most effective training method is the case study method followed by business games and role-playing. Fox, 1963 [26] had exposed some students to case study analysis and the outcome was: One-third of the students improved significantly to handle cases, one-third of the students showed moderate improvement in handling cases, and the remaining one-third showed no improvement in handling cases. Training in brainstorming could also improve problem-solving skills [27, 28]. Brainstorming can be taught by role-playing.

## 5.4 Enhancement of interpersonal skills/ people management skills

Interpersonal skills are goal-directed behaviors used in face-to-face communication to achieve desired objectives [29]. Interpersonal skills are also referred as people skills, face-to-face skills, social skills, and interpersonal ability. Sensitivity training is most effective, followed by role-playing, to address enhancement of interpersonal skills. While conference (discussion) method and case study method are also considered effective [30]. Role-playing is effective in increasing sensitivity to employee motivation [31]. Role-playing is also effective in improving interviewing skills [32]. Role-playing improves group leadership skills which are a form of interpersonal skills [33–36].

## 5.5 Retention of knowledge

Retention of knowledge acquired is the retrieval of information stored in long-term memory [37]. Programmed instruction is most effective followed by the case study method and the sensitivity training method. Role-playing, conference (discussion) method, and business games are significantly more effective training methods [38–41]. The study believes that the quantity of materials retained is usually proportional to the quantity learned [38, 39].

# 6 Training methods

Training methods taken into our scope of study in coherence to literature review are listed in Table 3.

**Table 3:** Training methods in our scope of study as per literature review.

| Sl. no. | Training methods | Definition | Source |
|---|---|---|---|
| 1 | Lecture method | Lecture method is a traditional method to illustrate information. It encompasses verbal presentation on a subject by a qualified person. | Reith [45], Golafrouz & Khaghanizadeh [43] |
| 2 | Programmed instruction method | Programmed instruction is a method of providing trainees or students with a graded series of guided steps for new subject matter. These trainees or students work by themselves at their own pace through the programmed material and test their understanding after each move by answering a question or filling in a diagram. | Ayutthaya & Koomsap [46]; Skinner [11, 14, 16] |
| 3 | Case study method | In case study method, the trainees are required to read and retain information independently from a particular case study; case studies are interactive in nature. | Martin et al. [48]; Shivakumar [50] |
| 4 | Sensitivity training method | Sensitivity training model is a group gathering without a plan to discuss topics relating to "here and now" of the group procedure. It denotes both a distinctive training method and specific content area. | Hinrichs [51]; Seashore [52] |
| 5 | Role-playing method | In role-playing an individual learns by watching, introspecting, correcting, or improvising the information obtained and by participating in the general assessment of the complete activity. | Sogunro [53]; Johnson & Johnson [54] |
| 6 | Business games method | Business games are a method of using experiential techniques to develop leaders. | Lopes et al. [64]; Faria [65] |

**Table 3** (continued)

| Sl. no. | Training methods | Definition | Source |
|---|---|---|---|
| 7 | Movie films method | Numerous research have indicated that videos embedded with training or education module have a huge impact in learning in the fields of psychology, counseling, medicine, management, and political science. | Borko and Pittman [68]; Brophy [69]; LeFevre [70]; Moreno & Valdez [71]; Pryor & Bitter [72]; Richardson & Kile [73]; Seago [74]; Sherin [75]; Wang & Hartley [76] |
| 8 | Conference (discussion) method | In conference (discussion) method, there is exchange of ideas, and these ideas are expanded, criticized, and honed. Also, for such discussions, the capacity to listen is an important skill. | Smith & Foley [82]; Costin [81] |
| 9 | Television lecture method | Television is a strong interaction tool with tremendous potential for informing, entertaining, and educating. Advanced mass communication technology presents teachers with the potential in the area of teaching and learning for a scientific revolution. | Jayakumar [84]; McIsaac & Guanwardena [86] |

***Lecture method:*** The lecture method was the only way to transfer knowledge from one generation to another before the industrial revolution, when books were not yet mass commodities [42]. One of the most common methods of training is the lecture method. Lecture method is an easy, quick, and inexpensive way to present the vast problems to many groups of learners [43]. However, the cons of lecture method also need to be looked up, such as long lectures, one-way communication which leads to forgetting of issues [44]. Lecture method is a traditional method to illustrate information. It encompasses verbal presentation on a subject by a qualified person [45]. Research depicts that lecture method is more effective when it comes to acquiring knowledge; however, learning is enhanced by 10% via programmed instruction in comparison to lecture method.

***Programmed instruction method:*** Programmed instruction is a method of providing trainees or students with a graded series of guided steps for new subject matter. These trainees or students work by themselves at their own pace through the programmed material and test their understanding after each move by answering a question or filling in a diagram [46]. New subject matter/knowledge/skill can be presented through programmed instruction method in a sequence of controlled steps. Trainees have the flexibility and space to go through the programmed materials at their own pace stepwise, and each step they are tested through examination question or filling a diagram. Instruction, whether aimed at developing academic skills in retarded children, increasing general knowledge in normal children; or advancing mathematical sophistication in gifted youngsters, may be conceived of as a process in which a teacher systematically and effectively arranges and rearranges an environment to bring about desired behavioral changes. Teaching on this basis, in whole or part, has been called as programmed instruction [15, 47]. Programmed instruction is a technology based on experimental analysis of behavior [11, 14, 16].

***Case study method:*** In case study method, the trainees are required to read and retain information independently from a particular case study; case studies are interactive in nature [48]. Case study method is the best method to improve problem-solving skills [49]. Case study helps in enhancing of decision-making skills, enhances communication and interpersonal skills, and also strengthens analytical skills. This training method is ideal for a situation where the learners have some prerequisite information and could benefit from the trainings essence of applicability. This is a low-risk, low-cost, training method that can concurrently train a large number of learners, because of which it is a popular choice for law, medicine, and counseling students training [50]. Carroll et al. (1972) [10] did a survey of industry-based trainers, and it was found that case study method was the second best method for knowledge acquisition; the best method was programmed instruction method. Also, for knowledge retention, case study method is the best training method after programmed instruction method as per the literature.

***Sensitivity training method:*** Sensitivity training method model is a group gathering without a plan to discuss topics relating to "here and now" of the group procedure [51]; it denotes both a distinctive training method and a specific content area. The objective of sensitivity training is classified into three categories: individual point of view, group point of view, and the organization point of view. Individual point of view: A trainee can get a clear picture of how he/she is perceived of any given act. Different individuals have different views of the same behavior. For example, to know the effect on any person as supportive or against it, meaningful or meaningless, clear or ambiguous. It is not very common for all components of a group to have the same view of any particular person or event [52].

***Role-playing method:*** In role-playing, an individual learns by watching, introspecting, correcting, or improvising the information obtained and by participating in the general assessment of the complete activity [53]. The more authentic the role-playing and more successful the practice, the more you will feel emotional engagement, and the more you will understand [54]. Role-playing as a training method involves playing a role as a method that is used in individual and group circumstances such as counseling, interviews, customer service and sales, private interactions, negotiations, government meetings, team committees, research, and procedures of change that involve a lot of practical involvement of the participants, in the work

[55–60]. Role-playing has the advantage of letting trainees actually practice relevant skills and thus has greater transferability to the workplace than do didactic lectures of abstract discussions about leadership [61]. Role-playing is particularly suitable for offering trainees the opportunity to practice and focus together on how to handle important responsibilities in their work, or for sensitivity training and other purposes that include attitude training [62]. Also, a research exploring the use of role-playing in education of genetic counseling students found that participants saw this approach as useful in educating them on the genetic counseling system and consequences, and this experience would help them interact with patients effectively [63].

***Business games method:*** Business games method is a process of using experiential techniques to develop leaders [64]. Business games are widely accepted in management education or training. Around 90% AACSB (Association to Advance Collegiate Schools of Business) schools are using business games as a management simulation for management education [65]. A review of 304 business simulation education and learning articles shows that the five aspects gained through business games are strategy, decision-making, teamwork, experience, and learning outcomes and objectives [66]. Business games as a training method epitomizes experiential learning [67]. A business game is a tool to test the understanding of theoretical insights through practical application.

***Movie films method:*** Numerous researches have indicated that videos embedded with training or education module have a huge impact in learning in the field of psychology, counseling, medicine, management, and political science [68–76]. Learning through pictures, that is, videos and audiovisual, is a superior learning as compared to verbal, that is, audio conditions. This is related to the picture superiority effect [77, 78]. In order to strengthen important concepts and ideas, video clips are used as they induce visual stimuli. Generally, information is often stored in visual form such as movie clips; pictures and movie clips can be very vital to help students remember crucial ideas and get back from their long-term memory [79]. Around 94% of adults in Malaysia were engaged in acquisition of knowledge by watching television, video, and films [80].

***Conference (discussion) method:*** In conference (discussion) method a trainee and a trainer holds one-to-one discussions on the subject matter to be imparted or taught. Discussion method is beneficial for speech courses for students with low ability; also it is a good technique that helps in recalling facts. Relatively discussion may be more effective method of teaching as compared to lecture method [81]. In conference (discussion) method, there is an exchange of ideas and these ideas are expanded, criticized, and honed. Also, for such discussions, the capacity to listen is an important skill [82]. According to Johnson and Smith, 1953 [83], students in discussion class in a basic psychology course not only expressed opinions about the course when compared to other students who were taught the same course via lecture method, but also, these students who studied through discussion or conference technique were more open to tests and grading procedures.

***Television lecture method:*** The television lectures are unique in a way as it includes communication between two or more locations where instructions are given at the moment specified (instructions that take place more than once at the same moment). Television is a strong interaction tool with tremendous potential for informing, entertaining, and educating. Advanced mass communication technology presents teachers with the potential in the area of teaching and learning for a scientific revolution [84]. Researchers have constantly discovered that television is capable of motivating and captivating learners and stimulating interest in the learning process [85]. Distance learning through interactive television lectures is one of the fastest-growing instructional medium [86]. Distance learning is not a new concept while videotaped lectures have been used for long in universities and professional courses for over two decades [87].

# 7 Methodology

A questionnaire was constructed on the latest effective training methods used to enhance learning of KWs in outsourcing organizations and administered to trainers, managers, senior managers, and learning and development employees who have been conducting training sessions or programs in outsourcing organizations. The questionnaire was designed in such a manner that for every training objective, three training methods were provided to respondents and they were asked to rank these three training methods for the specific training objective. These three training methods were derived from literature review, which is considered to be best for that specific training objective.

# 8 Use of Web 3.0 in training knowledge workers

Web 3.0 refers to a third wave of Internet-based services that collectively form what could be called "the intelligent web" [88]. In simple terms, Web 3.0 is indeed a technically advanced medium that allows users to read, write, and execute, and furthermore it enables machines to perform which earlier was expected from humans only [89]. It is needless to say that millions across the globe use Internet or the web to access information to learn. Web 3.0 is basically the third generation in web evolution. In order to understand the scope of Web 3.0 better we have to understand Web 1.0 and Web 2.0. Web 1.0 was a medium in which knowledge was built with text and photographs and could be published in static form. It represented an atmosphere in which data and knowledge were static and presented without interaction between the user and the information and limited content creators, also known as read-only web [90]. Web 2.0 is basically an expansion of Web 1.0's initial objectives, values, and underlying technology [91]. Web 2.0 has numerous benefits particularly the way in which it contributes to the development of complex knowledge networks by allowing users to engage in the web infrastructure itself, creating vast quantities of data with many possible uses. It has also contributed to the growth of user centric flexible applications [92].

Semantic web-based education (SWBE) is a platform to perform activities such as teaching, learning, collaboration, assessment, and other educational operations. Pedagogical representatives promote learning by engaging with learners and teachers and working with other representatives in an interactive learning atmosphere to enable content and information to flow. In SWBE, educational material focused on the web is described by learning objects. Learning objects are digital entities that can be reused for a learning resource such as electronic texts, interactive content, animations, and pictures [93].

***Web 3.0 in education and research:*** Web 2.0 learning emphasizes the active engagement of internet users and the connection between virtual communities through social networking tools or social applications such as blog, wiki, social book marking, and social networking. A more open approach to learning will be facilitated by resources and services of Web 3.0 technologies. Web 3.0 is a combination of high-powered graphics and textual data. According to ASTD (American Society for Training & Development) President and CEO Tony Bingham, content finds you on the semantic web instead of actively pursuing it. Your actions and preferences would then decide what you find and it will be presented to your channel of choice and how you would like it [89]. Various e-learning resources such as MOODLE, Web 2.0, Web 3.0, and Blackboard are available that can be used in higher education. E-learning tools can offer preparation and higher education to many students who come from diverse cultural backgrounds and have varying levels of higher education. E-learning 3.0 is one of the main trends for social learning using Web 3.0 tools. On the internet, it is also simple to quickly integrate new services into existing software such as integration of Wiki with Web 3.0. Web 3.0 has paved the evolution of e-learning and it entails lower costs relative to conventional learning and is seen as more efficient learning, hence facilitating diffusion of new processes and techniques which in turn reduces geographical challenges [94]. Personalization offers education plans customized to the participants based on an overview of the interests of the learners current skills and knowledge status, expectations for learning styles, as well as continuous process tracking [95]. The RDF, XML, and ontology-based Web 3.0 approach is ideal for personalizing learning objectives [96].

***Web 3.0 in learning and training in organizations:*** Web 3.0 has a lot of scope in organizational learning, yet it has not been used to its maximum potential yet. Web

3.0 innovations provide smarter searches with better and more focused performance and higher quality of knowledge acquisition in a shorter period. While the immersive interactive elements of Web 3.0 have been used for years in the field of learning in the corporate training and development sector with the use of virtual environments and simulations but other immersive technologies still remain underutilized. Individuals who are oriented on the role of Web 3.0 in the learning function and in IT correlates with both a good learning function and high market efficiency [97]. Some of the ways Web 3.0 technologies affect and influence learning today and in future primarily in high-performing organizations are:

- High-performing organizations are required to have divisions of learning and training open to new technologies and use these technologies to solve business problems.
- In the next few years, the use of virtual environments, simulations, augmented reality and multiplayer gaming technology for training in organizations is anticipated to grow substantially.
- It is time to be strategic and ready to transition with the seismic change from Web 2.0 to Web 3.0 for learning and training practitioners.

# 9 Findings and analysis

A questionnaire was designed and administered to knowledge workers working in the outsourcing industry. These knowledge workers comprised trainers, managers, senior managers, and learning and development professionals who had a fair amount of knowledge of these methods and their outcome. For every training objective, three training methods were provided to respondents and they were asked to rank these three training methods for the specific training objective. These three training methods were derived from literature review, which are considered to be best for that specific training objective.

For training objective "acquisition of new knowledge or skill," programmed instruction was ranked first followed by case study and conference (discussion) method. Training objectives such as change in attitude, according to respondents can be best addressed by sensitivity training followed by role playing and conference (discussion) method. Case study method was ranked first followed by business games and role-playing method to address training method problem-solving skills/troubleshooting skills. In order to address training objective enhancement of interpersonal skills/people management skills, sensitivity training method was ranked first while role-playing and conference (discussion) methods were ranked second and third respectively. The final training objective retention of knowledge can be best addressed by programmed instruction method followed by case study and sensitivity training method as per the respondents (Table 4).

**Table 4:** Training methods ranking as per respondents.

| Training objective | Training methods | Rank |
|---|---|---|
| Acquiring new knowledge or skill | Programmed instruction | 1 |
| | Case study | 2 |
| | Conference (discussion) | 3 |
| Change in attitude | Sensitivity training | 1 |
| | Role-playing | 2 |
| | Conference (discussion) | 3 |
| Problem-solving skills/troubleshooting skills | Case study | 1 |
| | Business games | 2 |
| | Role-playing | 3 |
| Enhancement of interpersonal skills/people management skills | Sensitivity training | 1 |
| | Role-playing | 2 |
| | Conference (discussion) | 3 |
| Retention of knowledge | Programmed instruction | 1 |
| | Case study | 2 |
| | Sensitivity training | 3 |

The respondents belonging to the age group of 25 to 34 years were 55.93%; 35 to 44 years 37.29%, 45 to 54 years 3.39%, and respondents under the age of 25 years were 3.39%. The classification and percentage of respondent's designation were: Trainer 25%, Training Manager 20%, Assistant Manager 16.67%, Team Leader 10%, Quality Manger 6.67%, General Manager 5%, Associate General Manager 5%, Operations Manager was 5%, Executive 5%, and Associate Vice President 1.67%. Respondents whose primary role is training were 58.33% followed by quality at 18.33%, and operations at 23.33%. From a qualification perspective maximum respondent's were graduates which was 60% followed by post graduates at 36.67% and undergraduates at 3.33%. Respondents who had more than 10 years of work experience constituted 61.67% and between 6 years to 10 years constituted 21.67%, and respondents who had work experience less than 5 years constituted 16.67%.

The respondents were asked, "How rigorous are their organization's training programmes?" The responses were: A great deal – 33.33%; A lot – 38.33%; A moderate amount – 18.33%; A little – 8.33%; and None at all – 1.67%. The respondents were asked to "List some of the Training Needs Identified in outsourcing organizations." The responses to this question were: Acquiring new knowledge or skill – 81.67%; Change in attitude – 43.33%; Problem-solving skills – 75.00%; Interpersonal skills – 53.33%; Knowledge retention – 53.33% and others were 10.00%. "Others" according to respondents constituted such as leadership training, client interaction and business communication, and resilience and customer centricity. The respondents opined that acquiring new knowledge or skill, problem-solving skills as critical training needs followed by interpersonal skills and knowledge retention.

A total of 48.33% of the respondents "strongly agree" that proper training methods are used to address the training needs that have been identified whereas 45% of the respondents "agree" that proper training methods are used to address the training needs that have been identified; and a total of 5% "disagree" that proper training methods are used to address the training needs that have been identified. While only 1.67% of the respondents "neither agree nor disagree." Role-playing method and programmed instruction method are widely used as training methods in the outsourcing services sector and their usage were denoted as 78.33% and 76.67% respectively; followed by case study method which is at 71.67%, conference (discussion) method at 66.67%, lecture method at 60%, business games at 55%, sensitivity training method at 53.33%, TV lecture method 45%, movie film method at 45%, and others (online portal) constituted only 1.67%.

# 10 Conclusion

The current training methods used in practice in the outsourcing services sector with respect to addressing a particular training need from a KW's perspective is somewhat similar to findings of Carroll et al. 1972 [10]. According to Carroll et al. 1972 [10], "Knowledge acquisition" could be best addressed by programmed instruction training method and our findings in the outsourcing services sector depict the same; that is, "acquisition of new knowledge or skill" could be best addressed by programmed instruction training method. "Change in attitude," according to Carroll et al. [10], could be best addressed by sensitivity training method, which is similar to our finding. Training need "problem solving skills," is best addressed by case study training method, followed by business games and role-playing method. "Knowledge retention" is best addressed by programmed instruction method followed by case study method and sensitivity training method.

Our findings state that role-playing training method is the most used training method in the outsourcing services sector. It is widely accepted and used to tackle training needs pertaining to "change in attitude" and "interpersonal skills." Programmed instruction training method is the second most popular training method in the outsourcing services sector. Case study training method is the third most popular training method in the outsourcing services sector. Case study as a training method was found to be the best to address training need "problem solving skill." Programmed instruction as a training method is considered to be best for the training need "acquiring new knowledge or skill and knowledge retention."

## 11 Scope for further study

Further studies can be specifically done on role-playing and business games training methods. There exists a very limited amount of literature when compared to other training methods. Also, role-playing has emerged as the second best training method in order to address training need "change in attitude" and "interpersonal skills" as per literature review. Also, our findings ranked role-playing training method as the third best training method to address training need "problem-solving skills" and second best training method to address "interpersonal skills." Business games as a training method is used in simulation exercises in managerial training and also it has its uses in academic training. Hence, further studies would definitely expand the body of knowledge pertaining to these two training methods.

This research received no specific grant from any funding agency in the public, commercial, or not-for-profit sectors.

## References

[1] Heathfield, S. (2008). "Twelve tips for team building: how to build successful work teams", available:http://humanresources.about.com/od/involvementteams/a/twelve_tip_team_3.htm (accessed 20th April 2019).

[2] (Mandakini, 2002) http://aladin.uil.unesco.org/paldin/pdf/course01/unit_10.pdf.

[3] Drucker, P.F. (1994). The age of social transformation. The Atlantic Monthly, 274(5), 53–80.

[4] Alvesson, M. (1995). Management of Knowledge-Intensive Companies, New York, Walter de Gruyter & Co.

[5] Maria, P.K. & Elena, G. (2008). Development of personalized learning objects for training adult educators of special groups. Journal of Knowledge Management, 12(6), 89–101.

[6] Arthur, J.R. (1996). An application of the high transfer Training Methodology with soft skills Task.

[7] Logan, J.M., Castel, A.D., Haber, S. & Viehman, E.J. (2012). Metacognition and the spacing effect: the role of repetition, feedback, and instruction on judgments of learning for massed and spaced rehearsal. Metacognition and Learning, 7(3), 175–195.

[8] Kolb, A.Y. & Kolb, D.A. (2005). Learning styles and learning spaces: Enhancing experiential learning in higher education. Academy of management learning & education, 4(2), 193–212.

[9] Sutton, M.A., Ide, J., Masters, S.E. & Carew, T.J. (2002). Interaction between amount and pattern of training in the induction of intermediate- and long-term memory for sensitization in Aplysia. Learning and Memory, 9, 29–40.

[10] Carroll, S.J., Paine, F.T. & Ivancevich, J.J. (1972). The Relative Effectiveness of Training Methods: Expert Opinion And Research, Personnel Psychology.

[11] Skinner, B.F. (1953). Science and Human Behavior, New York, NY, Free Press.

[12] Skinner, B.F. (1954). The science of learning and the art of teaching. Harvard Educational Review, 24, 86–97.

[13] Skinner, B.F. (1958). Teaching machines. Science, 128, 969–977.

[14] Skinner, B.F. (1961). Why we need teaching machines. Harvard Educational Review, 31, 377–398.

[15] Skinner, B.F. (1963). Operant behavior. American psychologist, 18(8), 503.
[16] Skinner, B.F. (1964). New methods and new aims in teaching. New scientist, 122(5), 483–484.
[17] Goguelin, P. (1960). Les problemes psychologiques dans l'industrie. Lecture at the Sorbonne on, 29(10).
[18] Petty, R.E. & Brinol, P. (2010). Attitude change. Advanced social psychology: The state of the science, 217–259.
[19] Festinger, L. & Carlsmith, J. (1959). Cognitive consequences of forced compliance. Journal of Abnormal and Social Psychology, 58, 203–210.
[20] Harvey, O.J. & Beverly, G. (1961). Some personality correlates of concept change through role playing. Journal of Abnormal and Social Psychology, 63, 125–130.
[21] Culbertson, F. (1957). Modification of an emotionally held attitude through role playing. Journal of Abnormal and Social Psychology, 54, 230–233.
[22] Janis, I. & King, B. (1954). The influence of role playing on opinion change. Journal of Abnormal and Social Psychology, 49, 211–218.
[23] King, B. & Janis, I. (1956). Comparison of the effectiveness of improvised vs. nonimprovised role playing in producing opinion changes. Human Relations, 9, 177–186.
[24] Janis, I. & Mann, L. (1965). Effectiveness of emotional role-playing in modifying smoking habits and attitudes. Journal of Experimental Research in Personality, 1, 84–90.
[25] Davidson, J.E., Sternberg, R.J. & Sternberg, R.J. (Eds.). (2003). The Psychology of Problem Solving, Cambridge university press.
[26] Fox, W.M. (1963). A measure of the effectiveness of the case method in teaching human relations. Personnel Administration, 26, 53–57.
[27] Parnes, S.J. & Meadow, A. (1959). Effects of brainstorming instructions on creative problem solving by trained and untrained subjects. Journal of EducationalPsychology, 50, 171–176.
[28] Cohen, D., Whitmyre, J.W. & Funk, W.H. (1960). Effect of group cohesiveness and training upon creative thinking. Journal of Applied Psychology, 44, 319–322.
[29] Hayes, J. (2002). Interpersonal Skills at Work, Routledge.
[30] Campbell, J.P. & Dunnette, M.D. (1968). Effectiveness of T-group experiences in managerial training and development. Psychological Bulletin, 70, 73–104.
[31] Bolda, R.A. & Lawshe, C.H. (1962). Evaluation of role playing. Personnel Administration, 25, 40–42.
[32] Van Schacck, H. Jr. (1957). Naturalistic role playing: A method of interview training for student personnel administrators. Dissertation Abstracts, 17, 801.
[33] Maier, N.R.F. (1953). An experimental test of the effect of training on discussion leadership. Human Relations, 6, 161–173.
[34] Maier, N.R.F. & Hoffman, L.R. (1960). Quality of first and second solutions in group problem solvipg. Journal of Applied Psychology, 44, 278–283.
[35] Maier, N.R.F. & Hoffman, L.R. (1960). Using trained "developmental" discussion leaders to improve further the quality of group decisions. Journal of Applied Psychology, 44, 247–251.
[36] Maier, N.R.F. & Maier, R.A. (1957). An experimental test of the effects of "developmental" vs. "free" discussions on the quality of group decisions. Journal of Applied Psychology, 41, 320–323.
[37] Bennett, A.G. & Rebello, N.S. (2012). Retention and Learning. In: Seel, N.M. (eds), Encyclopedia of the Sciences of Learning, Boston, MA, Springer.
[38] Dietrick, D.C. (1960). Review of research. In: Hill, R.J., A comparative study of lecture and discussion methods. Pasadena, California, The Fund for Adult Education, 90–118.
[39] Verner, C. & Dickinson, G. (1967). The lecture, an analysis and review of research. Adult Education, 17, 85–100.

[40] Sodnovitch, J.M. & Pophorn, W.J. (1961). Retention Value of Filmed Science Courses, Kansas State College of Pittsburg.

[41] VanderMeer, A.W. (1948). Relative Effectiveness of Exclusive Film Instruction, Films Plus Study Guides, and Typical Instructional Methods. Progress Report #lo. Instructional Film Program, State College, Pennsylvania, Pennsylvania State College.

[42] Mazur, E. (2009). Farewell, lecture? Science, 323(5910), 50–51.

[43] Golafrouz, S.H. & Khaghanizadeh, M. (2010). Introduction to oral presentation teaching method.

[44] Nowroozi, H.M., Mohsenizadeh, S.M., Jafari, S.H. & Ebrahimzadeh, S. (2011). The effect of teaching uning a blend of collaborative and mastery of learning models, on learning of vital signs: An experiment on nursing and operation room students of Mashhad University of Medical Sciences.

[45] Reith, J. (1976). Group methods: Conferences, meetings, workshops, seminars. In: Craig, R.L. (Ed.), Training and Development Handbook, 2nd Chapter 34, New York, McGraw-Hill.

[46] Ayutthaya, D.H.N. & Koomsap, P. (2019). LOVE-Based Teaching and Learning Method Classification. In To be appeared in Proceedings of INTED2019 Conference 11th-13th March.

[47] Weiss, R.J. & Green, E.J. (1962). The applicability of programmed instruction in a medical school curriculum. Academic Medicine, 37(8), 760–766.

[48] Martin, B.O., Kolomitro, K. & Lam, T.C. (2014). Training methods: A review and analysis. Human Resource Development Review, 13(1), 11–35.

[49] Mahoney, T.A., Jerdee, T.H. & Karman, A. (1960). An experimental evaluation of management development'. Personnel Psychology, 13(1), 81–98.

[50] Shivakumar, K. (2012). The case study method in training and management education. The IUP Journal of Soft Skills, VI(2), 55–64.

[51] Hinrichs, J.R. (1976). Personnel training. In: Dunnette, M.D. (Ed.), Handbook of Industrial and Organizational Psychology, Chicago, Rand McNally, 829–860.

[52] Seashore, C. (1968). What Is Sensitivity Training? News and Reports from NTL Institute for Behavioral Science, 2(2), 1–2.

[53] Sogunro, O.A. (2004). Efficacy of role-playing pedagogy in training leaders: some reflections. Journal of Management Development.

[54] Johnson, D.W. & Johnson, F.P. (1997). Joining Together: Group Theory and Group Skills, 6th, Boston, MA, Allyn & Bacon.

[55] Buchan, L.G. (1972). Roleplaying and the Educable Mentally Retarded, Belmont, CA, Fearon.

[56] Harris, B.M. (1985). Supervisory Behavior in Education, 3rd, Englewood Cliffs, NJ, Prentice-Hall.

[57] Hershey, G.L. (1985). Supervising Office Systems Personnel: Strategies for Professional Management of the Automated Office, Englewood Cliffs, NJ, Prentice Hall.

[58] Cranton, P. (1989). Planning Instruction for Adult Learners, Toronto, Wall & Emerson.

[59] Cranton, P. (1992). Working with Adult Learners, Toronto, Wall & Emerson.

[60] Ments, M.V. (1989). The Effective Use of Role-play: A Handbook for Teachers and Trainers, rev, London, Kogan Page.

[61] Hughes, R.L., Ginnett, R.C. & Curphy, G.J. (1993). Leadership: Enhancing the Lessons of Experience, Homewood, IL, Irwin.

[62] Barone, D.F., Hutchings, P.S., Kimmel, H.J., Traub, H.L., Cooper, J.T. & Marshall, C.M. (2005). Increasing empathic accuracy through practice and feedback in a clinical interviewing course. Journal of Social & Clinical Psychology, 24(2), 156–171.

[63] McIlvried, D.E., Prucka, S.K., Herbst, M., Barger, C. & Robin, N.H. (2008). The use of roleplay to enhance medical student understanding of genetic counseling. Genetic Medicine, 10, 739–744.

[64] Lopes, M.C., Fialho, F.A., Cunha, C.J. & Niveiros, S.I. (2013). Business games for leadership development: A systematic review. Simulation & Gaming, 44(4), 523–543.
[65] Faria, A.J. (1998). Business simulation games: Current usage levels—An update. Simulation & Gaming, 29, 295–308.
[66] Faria, A.J. Dr., Hutchinson, D. Dr., Wellington, W.J. Dr & Gold, S. (2009). Developments in Business Gaming A Review of the Past 40 Years. Simulation and Gaming, 40(4), 464–487.
[67] Garris, R., Ahlers, R. & Driskell, J.E. (2002). Games, motivation and learning: a research and practice model. Simulation & Gaming: An Interdisciplinary Journal, 33(4), 441–467.
[68] Borko, & Pittman, M.E. (2008). Video as a tool for fostering productive discussions in mathematics professional development. Teaching and Teacher Education, 24(2), 417–436.
[69] Brophy, J.E. (Ed.). (2004). Using video in Teacher Education, Oxford, UK, Elsevier.
[70] LeFevre, D.M. (2003). Designing for teacher learning: Video-based curriculum design. Advances in Research on Teaching, 10, 235–258.
[71] Moreno, R. & Valdez, A. (2007). Immediate and delayed effects of using a classroom case exemplar in teacher education: The role of presentation format. Journal of Educational Psychology, 99(1), 194–206.
[72] Pryor, C.R. & Bitter, G.G. (2008). Using multimedia to teach in service teachers: Impacts on learning, application, and retention. Computers in Human Behavior, 24(6), 2668–2681.
[73] Richardson, V. & Kile, R.S. (1999). The use of videocases in teacher education. In: Lundeberg, M.L., Levin, B. & Harrington, H. (Eds.), Who Learns from Cases and how? The Research base for Teaching with Cases, Mahwah, NJ, Erlbaum Associates, 121–136.
[74] Seago, N. (2003). Using video as an object of inquiry for mathematics teaching and learning. Advances in Research on Teaching, 10, 259–286.
[75] Sherin, M.G. (2003). New perspectives on the role of video in teacher education. Advances in Research on Teaching, 10, 1–27.
[76] Wang, J. & Hartley, K. (2003). Video technology as a support for teacher education reform. Journal of Technology and Teacher Education, 11(1), 105–138.
[77] Nelson, D.L., Reed, V.S. & Walling, J.R. (1976). Pictorial superiority effect. Journal of Experimental Psychology. Human Learning and Memory, 2, 523–528.
[78] Paivio, A., Rogers, T.B. & Smythe, P.C. (1968). Why are pictures easier to recall than words? Psychonomic Science, 11, 137–138.
[79] Sexton, R.L. (2006). Using short movie and television clips in the economics principles class. The Journal of Economic Education, 37(4), 406–417. doi: 10.3200/jece.37.4.406-417.
[80] Kamis, M., Muhamad, M., Junoh, A.M., Asmuni, A. & Idris, K. (2005). Informal learning in Malaysia. Proceedings of the 46th Annual Adult Education Research Conference, University of Georgia, Athens.
[81] Costin, F. (1972). Lecturing versus other methods of teaching: A review of research. British journal of educational technology, 3(1), 4–31.
[82] Smith, L.A. & Foley, M. (2009). Partners in a human enterprise: Harkness teaching in the history classroom. The History Teacher, 42(4), 477–496.
[83] Johnson, D.M. & Smith, M.C. (1953). Democratic leadership in the college classroom. Psychological Monographs, 67(11), (Whole No. 361).
[84] Jayakumar, R. (2016). Opinion of the university teachers towards educational television programmes. American Journal of Education and Learning, 1(1), 45–52.
[85] Sherry, L. (1996). Issues in distance learning. International Journal of Educational Tele- communications, 1, 337–365.
[86] McIsaac, M. & Guanwardena, C. (1996). Distance education. In: Jonassen, D. (Ed.), Handbook of Research for Educational Communication and Technology, New York, Simon & Schuster Macmillan, 403–437.

[87] Moore, D.R. & Lockee, B.B. (1998). A taxonomy of bandwidth: Considerations and principles to guide practice in the design and delivery of distance education. Unpublished manuscript. Portland State University.
[88] Markoff, J. (2006). Entrepreneurs see a web guided by common sense. New York Times, 12.
[89] Lal, M. (2011). Web 3.0 in Education & Research. BVICAM's International Journal of Information Technology, 3, 2.
[90] Rudman, R.J. (2010). Incremental negative impacts in Web 2.0 applications. The Electronic Library, 28(2), 210–230.
[91] Anderson, P. (2007). All That Glisters Is Not Gold'—Web 2.0 And The Librarian.
[92] Miranda, P., Isaias, P. & Costa, C.J. (2014). E-Learning and web generations: Towards Web 3.0 and E-Learning 3.0. International Proceedings of Economics Development and Research, 81, 92.
[93] Devedzic, V. (2006). Semantic Web and Education (Integrated Series in Information Systems), 1st edn, New York, Springer.
[94] Aljawarneh, S.A. (2020). Reviewing and exploring innovative ubiquitous learning tools in higher education. Journal of computing in higher education, 32(1), 57–73.
[95] Sampson, D. (2002). Current and future research and technology developments in elearning. Industry and Higher Education Journal, 16, 4.
[96] Kurilovas, E., Kubilinskiene, S. & Dagiene, V. (2014). Web 3.0–Based personalisation of learning objects in virtual learning environments. Computers in Human Behavior, 30, 654–662.
[97] Green, M. (2011). Better, smarter, faster: Web 3.0 and the future of learning. Development and Learning in Organizations: An International Journal.

# Index



https://doi.org/10.1515/9783110722789-012